HANDLING RADIOACTIVITY

Handling Radioactivity

A PRACTICAL APPROACH FOR SCIENTISTS AND ENGINEERS

DONALD C. STEWART

Associate Director (Retired)
Chemistry Division
Argonne National Laboratory

A WILEY-INTERSCIENCE PUBLICATION

JOHN WILEY & SONS

New York • Chichester • Brisbane • Toronto

Library of Congress Cataloging in Publication Data:

Stewart, Donald Charles, 1912–
    Handling radioactivity.

    "A Wiley-Interscience publication."
    Bibliography: p.
    Includes index.
    1.   Radioactive substances—Safety measures.
I. Title.
TK9151.4.S83   621.48′37′0289   80-19258
ISBN 0-471-04557-8

Printed in the United States of America

10 9 8 7 6 5 4 3 2 1

To the three women in my life—
Dorothy, Kate and Deb

Preface

The aim of this book is to present an overall view in a descriptive and essentially nonmathematical way of the practicalities of handling radioactivity. It is hoped that the material will be particularly helpful to those entering the nuclear field for the first time and to those working in related areas whose responsibilities require them to have a general knowledge of the subject of radioactivity handling and its vocabulary. There may also be others like myself who in spite of thirty-six years experience in the field found during the preparation of the book that my knowledge of some of the subjects was less than perfect.

The presentation is primarily for bench-scale operations and with a bent that reflects my own background as a chemist. There is a considerable emphasis on facilities since these are fundamental to the safe handling of active materials. Facility design and detail is also unfortunately an area where the relevant information is largely scattered through literature sources that are not accessible to most readers.

Some of the topics surveyed—such as dosimetry, shielding and nuclear criticality—are extremely complex and no pretense is made that the treatment here represents more than bare bone summaries of the fields. A considerable effort has however been made to cite the key references in each area where more detailed information can be found. A few additional useful references not cited directly in the text appear in an abbreviated bibliography at the end of the book.

Many people helped by reviewing parts of the manuscript and answering innumerable questions. I should like to express my appreciation to many of my old associates at Argonne National Laboratory among whom were J. G. Ello, S. M. Fried, Sheffield Gordon, W. V. Lipton, F. P. Marchetti, H. J. Moe, Isabella Anderson, M. C. Sauer, Jacob Sedlet, E. P. Steinberg, J. H. Talboy and N. P. Zaichek; with special thanks to two old comrades-in-arms in the hot lab wars, S. J. Vachta and C. H. Youngquist. I am also grateful for help given by P. S. Baker and J. W. Gollehon of Oak Ridge, A. J. Branton of the British Department of Education and Science, Richard Harris of the American Nuclear Society, G. W. Köhler of the Gesellschaft für Kernforschung at Karlsruhe, and W. H. Schulze and Thomas Esit, both of the Chicago office of the Nuclear Regulatory Commission.

D. C. STEWART

Downer's Grove, Ill.
January 1981

Contents

Tables

HANDLING RADIOACTIVITY

chapter 1

Introduction and Background

Radioactive materials (chiefly radium) were finding minor uses in medicine and in industry in the period immediately prior to World War II on the milligram scale or less, usually much less. Now, forty years later, radioisotopes are being routinely processed in multikilogram quantities in many parts of the world. Obviously much has been learned about radioactivity and its handling in the interim as evidenced by the overwhelming volume of the nuclear energy literature.

This book is however not concerned with kilogram-quantity, industrial-level operations (although there are overlaps) but with the practical aspects of safely handling radioisotopes in the laboratory or on an equivalent scale in a shielded cell. A considerable emphasis has been placed on facility design and operation since these factors are fundamental to safe handling of what are admittedly hazardous materials. Working with radioactivity however presents no more threat, and possibly less, than do many other aspects of our technological civilization if a good facility, sensible operating rules, and a knowledgable staff are all present.

The topics surveyed are many in number and almost uniformly complex. Some basic knowledge of nuclear terms and processes on the part of the reader has therefore been assumed in order to avoid lengthy definitions and to keep the book length to a reasonable size.

1.1 NOTES ON FORMAT

Today's literature, scientific or otherwise, is filled with abbreviations and acronyms, a phenomenon that seems to be particularly prevalent in the nuclear energy field. Cryptic assortments eof upper-case letters will accordingly appear all through this book (and particularly in the next section). Each term will be defined at its first appearance and space is also reserved in the appendix for a recapitulation.

Many of the references given at the end of each chapter were published by an organization popularly referred to by its abbreviation (e.g., AEC). Again, the full name (and other possibly useful information if the group is still in existence) will be given at the first encounter, but only the name or an abbreviated form will be used in later citations in order to avoid continuous repetition. A number of the sources are cited in more than one chapter. When this occurs the later appearances will be cross-referenced back to the original.

1.2 THE NUCLEAR LITERATURE

When *Nuclear Science Abstracts* (NSA) issued its first volume in 1948 it carried 2023 entries abstracting the papers and reports in the nuclear field that had appeared over a six-month period. The final volume (No. 33) of the journal surveyed January to June of 1976 and contained 33,604 entries. The information explosion thus demonstrated is of course not unique to nuclear energy, but the area certainly has not been slighted.

The problem this presents to the individual attempting to keep abreast of developments in even a small area of specialization is obvious. The abstract journals help considerably and perform a particularly useful function in the nuclear energy field where much of the current information appears in organizational reports that are not readily available or in papers presented at meetings whose proceedings may not appear until some time later, if at all. Because of the steadily increasing volume of such materials the abstract journals will be eventually all probably replaced by computer searches, a process already well under way.

Nuclear Science Abstracts was originally issued by the Division of Technical Information of the U.S. Atomic Energy Commission (AEC). When that agency disappeared from the scene on January 17, 1975, its short-lived successor, the Energy Research and Development Administration (ERDA) continued NSA briefly while simultaneously initiating publication of *ERDA Energy Research Abstracts* (EERA). When ERDA itself was absorbed into the new Department of Energy (DOE) on October 1, 1977, EERA changed to ERA or simply *Energy Research Abstracts*.[1] This journal reviews material in all areas of energy research, meaning a concomitant less complete coverage of the nuclear literature. Fortunately this last review function has been assumed by the *Atomindex* of the International Nuclear Information System (INIS),[2] an abstract service that was started by the International Atomic Energy Agency (IAEA) in 1970. The first six volumes of *Atomindex* are available largely only on microfiche, but subsequent issues now appear semimonthly in journal form. The areas covered are very similar to those formerly surveyed by NSA (coverage of overseas work in probably even more comprehensive). An occasional abstract may appear in French, Spanish, or Russian, but most are in English. *Chemical Abstracts*, issued by the American Chemical Society,[3] surveys a more restricted number of papers in the basic nuclear science field, primarily those published in journal form.

The NSA entries were digitized in standard format for computer retrieval beginning with Volume 21 in 1967. As a result of the AEC to ERDA to DOE transition, this data base is now administrated by the DOE Technical Information Center (TIC) at Oak Ridge,[1] the successor to the former AEC Division of Technical Information Division Extension at the same location. The data base has now been considerably broadened to include reviews of all the various energy fields in addition to nuclear; it now encompasses well over 1,200,000 items searchable by computer with additions at the rate of 165,000 per year. Only the citation and subject indexing were initially computer retrievable, but complete abstracts were added in 1976. This is unquestionably the largest collecion of references on energy-related research and development in existence. New material is put onto magnetic tapes as it is acquired, appropriate selections then being submitted to to INIS *Atomindex* on an exchange basis at the rate of about 20,000 per year. Tapes are also sent to the National Technical Information Service[4] to be made available to the public. Tapes are also sent directly to information centers in other countries if there are existing exchange agreements.

The Technical Information Center is a part of the Department of Energy and as such receives copies of all printed reports issued by DOE or its contractors. TIC also acquires similar reports, patents, bibliographies, etc., from other agencies and foreign countries as well as reviewing each issue of some 900 scientific and technical journals for energy-related papers. Many of the internal reports are redistributed to DOE contractors, additional copies being printed at Oak Ridge if necessary.[1a] Much of the material however goes out as microfiche rather than in printed form, the microfiches being prepared by a commercial processor[5] devoted exclusively to DOE interests. Members of the public (save booksellers and cover organizations) can purchase reports from this source in microfiche form. Copies are also sent to the National Technical Information Service where they are reproduced for sale as duplicate negatives or as enlarged hard copy.

On-line computer access to the DOE data bases is limited to DOE and its contractors through the DOE/RECON (REmote CONsole) system which consists of forty dedicated terminals on three telephone circuits (east coast, west coast, and local). In addition there are approximately 180 dial-up terminals accessing the RECON system. Non-DOE users may, for a fee, obtain access through the Western Regional Information Center or through the DOE information centers in San Francisco and Denver.[1b]

The Technical Information Center also publishes an annual summary[1c] of the technical books and monographs sponsored by DOE. This catalogue of course now covers energy areas other than strictly nuclear.

AEC-ERDA-DOE and other agencies have also from time to time established specialized information centers. The 1974 edition of the Directory of Information Sources in the United States,[6] issued by the Library of Congress, lists some sixteen such organizations in the nuclear science area, those most pertinent to this book being the Bureau of Radiological Health of the Food and Drug Administration;[7] the Radiation Chemistry Data Center at the University of Notre Dame;[8] the Gamma-ray Spectrum Catalogue at the National Reactor Testing Station;[9] the National Neutron Cross Section Center at Brookhaven National Laboratory;[10] and the Nuclear Data Project, the Nuclear Safety Information Center, the Criticality Data Center, and the Radiation Shielding Information Center, all at Oak Ridge.[11]

Many scientific and technological journals have come into the nuclear field over the years, some of which have unfortunately also gone. (*Nucleonics*, puclished by McGraw-Hill, and *Isotopes and Radiation Technology*, sponsored by the AEC, were two such casualties in which much practical information of the type of interest here was published.) Those publications still in existence are too numerous to list completely. Many of the fundamental nuclear science papers appear in the physics journals such as *Physical Review* and *Nuclear Physics*. Basic chemistry- and biology-oriented papers are in the *Journal of Inorganic and Nuclear Chemistry*, or in some of the more specialized journals such as *Health Physics, Radiochimica Acta, Radiation Effects, Journal of Nuclear Medicine, Nuclear Safety, Journal of Radioanalytical Chemistry*, etc. The IAEA issues the *Atomic Energy Review* and many other specialized publications, a number of which will be cited in later chapters. *Nuclear Data Tables* is produced by Academic Press. Occasional papers directly pertinent to nuclear science also appear in less specialized publications such as *Science, Nature, Journal of the American Chemical Society, Analytical Chemistry* and in the general physics, metallurgy, biology and engineering literature.

Since equipment and hardware of various types are discussed in later chapters, mention should be made of sources of suppliers of such items. The best of these for present purposes is the *Buyer's Guide* published annually as a special issue of the American Nuclear Society's *Nuclear News*.[12] The *Guide to Scientific Instruments* issued by the American Association for the Advancement of Science[13] and the *Lab-Guide* published by the American Chemical Society,[14] while more general, give good coverage of nuclear specialities.

The difficulty of organizing such a wealth of information has of course necessitated many review articles and other conscious efforts to bring material together in specialized areas. There are a number of *Advances in* series (mostly on an annual basis) in a number of different fields: *Activation Analysis, Inorganic Chemistry and Radiochemistry, Nuclear Physics, Nuclear Science and Technology, Radiation Biology, Radiation Chemistry*, etc. *Progress in Nuclear Energy*, produced by Pergamon Press,[15] consists of twelve different series of books, each of long standing, on nuclear topics ranging from Physics and Mathematics to Law and Administration. The issues in each of the series tend to come forth on an irregular basis as the respective editors decide upon the new advances that should be reviewed.

A number of handbooks and multivolume compendia have also been published in specific areas, many of which will be cited in later chapters. A good deal of the experience gained under high-pressure conditions during the Manhattan Project period, 1942–1946, is still very valid (the laws of physics and chemistry presumably not having changed very rapidly during the interim) and is published in the *National Nuclear Energy Series*,[16] an older but still very fundamental source of information. The proceedings of the various "Atoms for Peace," United Nations sponsored meetings in Geneva, Switzerland, will be referred to in later citations. The review papers presented there (at least on the part of the United States) are very good and candid summaries of experience up to the time of their presentation. On a more public-information level, but well written and useful, not only for general education purposes but also for those already in the fold, was the *Understanding the Atom* series of pamphlets originally issued by the AEC, but now unfortunately yet another victim of the budget wars.[1]

Many other organizations such as the American Nuclear Society have extensive publication programs and will be cited.

REFERENCES

1 U.S. Department of Energy (DOE), Technical Information Center (TIC), P. O. Box X, Oak Ridge, TN 37830. Queries concerning *Nuclear Science Abstracts*, *ERDA Energy Research Abstracts*, *Energy Research Abstracts*, and the *Understanding the Atom* series should be submitted to the Center. (a) Anon, "DOE Technical Information Center, Its Functions and Services." USDOE Report TID-4660-R2 (1978); (b) Anon, "Energy Information Bases." USDOE Report TID-22783 (1978); (c) Anon, "Technical Reports and Monographs, 1978 Catalogue." USDOE Report TID-4582-R13 (1979).

2 International Atomic Energy Agency (IAEA), International Nuclear Information System (INIS), "*Atomindex*." Queries regarding this abstract journal and catalogue requests for other IAEA publications should be submitted to IAEA, Division of Publications, P. O. Box 100, A-1400, Vienna, Austria, or to UNIPUB, P. O. Box 433, Murray Hill Station, New York, NY 10016.

3 American Chemical Society (ACS), "*Chemical Abstracts*." ACS, 1155 Sixteenth Street, N. W., Washington, D.C. 20036.

4 National Technical Information Service (NTIS), 5285 Port Royal Road, Springfield, Va. 22151.

5 Magnagard, Inc., P. O. Box 3501, Oak Ridge, TN 37830.

6 Library of Congress, "A Directory of Information Sources in the United States: Federal Government." U.S. Government Printing Office, North Capitol and H Streets, N.W., Washington, D.C. 20401 (1974).

7 Bureau of Radiological Health, U.S. Food and Drug Administration, 5600 Fisher's Lane, Rockville, MD 20852.

8 Radiation Chemistry Data Center, Radiation Center, University of Notre Dame, Notre Dame, IN 46556.

9 Gamma-ray Spectrum Catalogue, National Reactor Testing Station, Idaho Falls, ID 83401.

10 National Neutron Cross Section Center, Brookhaven National Laboratory, Upton, NY 11973.

11 The Nuclear Data Project and the Radiation Shielding Information Center are at Oak Ridge National Laboratory, Box X, Oak Ridge, TN 37830. Enquiries to the Criticality Data Center and to the Nuclear Safety Information Center should be addressed to Box Y at the same address.

12 American Nuclear Society (ANS), "Buyer's Guide, 1979," a special issue of *Nuclear News*, a monthly publication. ANS, 555 N. Kensington Ave., La Grange Park, IL 60525.

13 American Association for the Advancement of Science (AAAS), "Science Guide to Scientific Instruments, 1979-1980." AAAS, 1515 Massachusetts Ave., Washington, D.C. 20036.

14 American Chemical Society (Ref. 3 above), "Lab-Guide, 1976-1977."

15 "Progress in Nuclear Energy, Series I-XII."Pergamon Press, Ltd., Headington Hill Hall, Oxford, England; Pergamon Press, Inc., Maxwell House, Fairview Park, Elmsford, NY 10523.

16 "National Nuclear Energy Series." Published 1948-1956 in ten divisions, approximately sixty volumes total. McGraw-Hill Book Co., 330 W. 42nd Street, New York, NY 10036.

chapter 2

Radiation Protection Standards

The Grecian goddess Athena was reputed to have sprung fully grown and armored from the head of Jupiter. To many people, today's problems of radiation protection would appear to fall in the same category, dating back to the discovery of nuclear fission in 1939 and its subsequent application during the Manhattan Project in World War II. This is not quite the case. The full-bodied man-made radiation imp was there before the beginning of the century, although admittedly smaller in size. In fact, he had been around *au naturel* since Genesis. The exposure received by the average person due to naturally occurring radioactivity (the uranium and thorium decay chains, potassium-40, cosmic rays and various other minor contributors) is still appreciably higher than from any other source and over a thousand times larger than the average contribution from any part of the nuclear fuel cycle.

The estimates of the exposure of the population at large to ionizing radiation vary somewhat, depending on the assumptions made by the estimators, but the ordering of sources is always the same even if there are some differences in the magnitude of the numbers. The Environmental Protection Agency (EPA) has presented[1] their evaluation of the U.S. situation as they calculate it to be in 1980:

Source	Average Exposure Per U.S. Individual (mrem[a]/yr)
Natural radiation	130
Medical applications	86
Nuclear bomb fallout	5
Industrial applications and Miscellaneous	0.3
Nuclear energy cycle	0.1

[a]Defined later in the chapter

7

Medical exposures account for 90%–95% of the man-made exposures to ionizing radiation, which in turn is only 70% of the level that the average man has received twenty-four hours per day for aeons. The term "medical applications" includes medical and dental diagnostic x-ray procedures, nuclear medicine applications, and therapeutic treatments with various types of radiation; the total being averaged over the whole population. It is of course obvious that some individuals receive much more, and others correspondingly less than the 86-mrem/yr figure shown. The above-average yearly exposures received by radiologists and dentists and their assistants are counted in as part of the "miscellaneous" total.

The early awareness of the hazards associated with the powerful new tools made available to the medical profession by controllable radiation is the background for the development of the various nongovernmental and govermental agencies concerned with protection measures as presented in the following sections. Control of medically oriented exposure is a continuing concern for many of the regulatory agencies for reasons that are obvious from the EPA tabulation.

2.1 THE INTERNATIONAL COMMISSIONS AND NCRP

The first few months of 1896 were truly remarkable from the point of view of man's understanding of the workings of the world around him and how that knowledge might be usefully applied. In early January, Roentgen announced his discovery of x-rays, and in February, Becquerel reported his evidence for the existence of "rays" emanating from uranium ores, that is, the demonstration of naturally occurring radioactivity and of the fact that matter was not immutable. Returning to January, it was only twenty-five days after Roentgen's announcement that Grubbe, a manufacturer and experimenter with Crooke's tubes (which emitted x-rays when accelerated electrons struck a metallic target) utilized one of his devices for treatment of a patient with carcinoma of the breast. He thus became the first person to apply x-rays for therapeutic purposes and could claim to be the first of the radiologists. In addition, since he used lead sheeting to shield the unexposed parts of the patient's body, he could also qualify as the first health physicist. [His belief in the ability of x-rays to destroy tissue was well-founded on unfortunate personal experience. During his experiments with cathode-ray tubes he had developed redness on the back of one hand (erythema) which quickly reached the blistering, skin-cracking and finally very painful ulcerating stages. Grubbe had had to seek medical help for his own obviously serious problems only a few days before attempting his cancer-killing experiment.[2]]

Becquerel's discovery of natural radioactivity created a great deal of excitement among the knowledgable, and by 1904 Marie Curie had isolated 100 mg of pure radium-226 from two tons of uranium pitchblende ore. Thus the medical profession had two powerful new tools available—x-rays and radium. Over the next ten to twenty years both became essential for certain types of diagnosis and therapy (and were abused by quacks). Some industrial uses (luminescent watch dials, etc.) also developed.

Each of the earlier practitioners of radiation applications had to develop his own techniques, largely by trial and error. Exposure times were essentially on an experimental basis since there was no quantitative way for one radiologist to compare his experience with that of another, and understanding of the mechanisms whereby radioactivity reacted with tissue was virtually nonexistent. An empirical measurement was tried for a time as an effort towards standardization—the "erythema dose" [apparently of the order of 500–1000 R (roentgens) of x-ray exposure in modern terms][3]—based on the amount of radiation producing visible reddening of the skin. Since this response differed from person to person, the unit was far from satisfactory. Patient and operator protection still largely remained an experimental process and one with unforunate consequences for everyone concerned if the guessing was bad. This obviously was not a good situation, and by the early 1920s there was a definite movement for better standards and stricter control of radiation applications.

Concern for radiation protection was felt in many countries and had much to do with the convening of the First International Congress of Radiology in 1925. One of the major actions of the Congress was to establish the International Commission on Radiological Units and Measurements[4] (ICRU), one of whose charges was to develop a radiation-exposure unit acceptable to everyone. (There apparently was also some discussion of an exposure limit of 0.1 erythema dose per year, but it is not clear that any official action was taken.[3]) The ICRU by the time of the Second Congress in 1928 had defined the roentgen as an exposure unit for x-rays, a definition that was accepted by the Congress. The roentgen will be discussed further later in the chapter.

While the question of an exposure unit was being attacked there still remained the problem (of lesser importance to many at the time) of defining "safe" limits of exposure, i.e., the establishment of radiation protection standards. The Second Congress accordingly adopted some rather general standards that had been in use in Great Britain for several years and also established the International X-Ray and Radium Protection Committee.[3] This group recommended an exposure limit of 0.2 R/day in 1934, but essentially stopped functioning (as did ICRU) from about 1937 until after World War II. In 1950 the X-Ray Committee was reorganized as the International Commission on Radiological Protection (ICRP). Both Commissions have been very active and influential since that time. (The ICRP operates through five committees, each of which publishes its recommendations separately.)

While the ICRU and ICRP are international in scope, parallel national organizations developed in this country and in others, a fortunate fact in view of the hiatus caused in world-wide activities by World War II. In the United States the American Standards Association was involved at an early date, primarily in setting standards for industrial protection in using x-ray equipment and radium.[5] A more broadly oriented organization was also established in

1929 under the sponsorship of the National Bureau of Standards (NBS): the Advisory Committee on X-Ray and Radium Protection,[6] originally composed of representatives of certain medical groups and equipment manufacturers. The tremendous expansion of radiation-related problems brought on by the advent of nuclear energy caused the committee to be much broadened in both numbers and responsibility in 1946 when the name was changed to the National Committee on Radiation Protection. This became the National Council on Radiation Protection (NCRP) in 1964 when the group was reorganized as a nonprofit corporation. Partly because of the disruptions caused elsewhere by the war, the NCRP has been influential on a much broader than a purely national scale. They produced the first true x-ray protection rules in 1931, dropped the maximum permissible dose (MPD) to 0.1 R/day in 1935 because of evidence that more than skin burns was involved in overexposure, and lowered the MPD once more in 1947 to 0.3 R/week because of additional evidence of possible genetic effects. The NCRP recommendations were essentially those adopted by the ICRP when it was rejuvenated in 1950.[3]

The NCRP operates through some fifty-four separate scientific subcommittees, each responsible for reviewing data and formulating recommendations in a particular area.[7] This elaborate structure developed due to an increasingly broader view of the question of radiation protection. Lasers, TV sets, microwave ovens, high-voltage transmission lines, and many other artifacts of our civilization now come in for examination by the NCRP through the various subcommittees.

The earlier NCRP reports were published as NBS handbooks as were some of the first ICRU recommendations. Both organizations now publish their materials directly. The ICRP reports are usually issued in hardback by Pergamon Press. All three groups are frequently cited throughout this book and selected lists from the publications of each are given in the Appendix.

2.2   THE INTERNATIONAL ATOMIC ENERGY AGENCY

The IAEA was established in 1957 as an automonous agency under the aegis of the United Nations and at present includes 100 countries in its membership. Among the many activities of the agency are included the dissemination of radiation protection information through a continuing and effective program of standards establishment, sponsorship of conferences and publication of handbooks, a number of which are cited throughout this book.

2.3   THE BEIR AND UNSCEAR REPORTS

Two other nongovernmental groups deserve special mention because of the widespread influence of their reports. Both are primarily concerned with the more subtle effects of radiation rather than in specifying formal standards or operating procedures. The conclusions of both groups however carry much weight and generate much heated discussion in radiation protection circles.

The first of these committee reports originates with the Advisory Committee on the Biological Effects of Ionizing Radiation of the National Academy of Sciences-National Research Council (the BEIR reports).[8] The second organization is the Scientific Committee on the Effects of Atomic Radiation of the United Nations (UNSCEAR).[9] A question of major concern to both groups is the existence or nonexistence of a threshold for biological damage from radiation. At very low radiation levels any effects that occur are extremetly difficult to separate experimentally from those arising from individual differences and the effects of natural background and the other stresses to which all living beings are subject. The question is then whether or not damaging results seen in high-level exposures to radiation can be extrapolated back to the low-level situation, particularly at the gene-damage level. The threshold proponents argue no—the body's repair mechanisms can maintain the status quo when the exposure level is very low.

2.4 OTHER NONGOVERNMENTAL GROUPS

A number of professional societies and specialized journals have naturally also developed in the radiation protection field, most of them rather medically oriented as would be expected from the nature of the subject and the historical background sketched above. These organizations and publications are too numerous to be listed here, but are given in a "Bookshelf on Radiological Health" originating from the Department of Health, Education and Welfare (HEW).[10]

Essentially all of the organizations discussed above are nongovernmental or only quasigovernmental, so while they can recommend, they cannot regulate. To a remarkable extent, however, their recommendations have formed the basis for most of the rules established to date around the world for control of radiation hazards. The political mechanisms for promulgating and enforcing these regulations of course vary from country to country, although a national Atomic Energy Commission is a commonly used political structure for assignment of the responsibility. A brief survey of the U.S. situation will be attempted, although with some trepidation since there are numerous groups involved and the precise extent of the responsibilities and authority of each is not always clear.

2.5 U.S. REGULATORY AGENCIES

Taylor[5] has reported on the American situation as it existed in 1955 in a paper presented at the first "Atoms for Peace" meeting in Geneva. Prior to 1950 there were essentially no formal regulatory codes relating to radiation existent in the United States, any court cases involving radioactivity being settled by reference to the unofficial standards in the NCRP handbooks. The Food and Drug Administration (FDA) did require that labels warning of hazard be affixed to certain types of x-ray machines. The Atomic Energy Commission (AEC) had been established in 1946, but the enabling Act was imprecise on the matter of promulgating radiation protection standards, so the agency had taken no action. The Atomic Energy Act of 1954 was more definite, so the AEC was engaged in formulating some rules at the time of the Geneva meeting. California had some radiation control regulations, but no procedures or staff for licensing and inspection. In 1955 the State of New York had a set of regulations about ready to go and New York City had a few limited rules regarding x-ray equipment. Pennsylvania, Massachusetts, West Virginia, New Jersey, Michigan and a few other states were at various stages of considering the need for some sort of legislative control within their borders. (While recognizing that more regulation was needed and inevitable, there is a rather wistful hope apparent in Dr. Taylor's paper to the effect that any new rules be kept to a minimum and as simple as possible. It was a nice thought.)

To return to the AEC, its mandate was far from being universal even after it was clear that the agency did have a standards-setting responsibility. Its area of influence was restricted to "by-product material, source material and special nuclear material", that is, they had no jurisdiction over x-ray machines, accelerator-produced isotopes, the accelerators themselves (unless they were AEC sponsored), naturally occurring radioisotopes (save source materials) or the multitude of devices producing nonionizing but nevertheless potentially hazardous radiation.

[By-product material is ". . .any radioactive material (except special nuclear material) yielded in or made radioactive by exposure to radiation incident to the process of producing or utilizing special nuclear material,"[11] that is, fission or activation products resulting from reactor operation. Source material is ". . .(1) uranium, thorium, or any combination thereof in any physical or chemical form, or (2) ores that contain 0.05 wt% or more of uranium, thorium, or any combination thereof."[12] Special nuclear material means ". . .(1) plutonium, uranium-233, uranium enriched in the isotope 233 or in the isotope 235, or any other material that the (Commission) determines to be special material, but does not include source material; or (2) any material artificially enriched by any of the foregoing, but does not include source. material."[13] In the business, incidentally, this last material is referred to as SNM.]

The AEC was eliminated as of January 19, 1975. Its regulatory responsibilities were assigned to the newly formed Nuclear Regulatory Commission, while its research and development functions in the atomic energy field became part of the equally new Energy Research and Development Administration (ERDA), which in turn was largely absorbed into the Department of Energy (DOE) as of October 1, 1977. The general radiation responsibility of course remains with the Nuclear Regulatory Commission (NRC), although DOE retains the authority in relation to nuclear energy research and development activities within its own organization and those conducted by its contractors, including the National Laboratories.

The regulations originally established by the AEC and now taken over by NRC appear in the Code of Federal Regulations, Title 10. Chapter 1—Energy. The "Title" is divided into "Parts"; those most directly pertinent to this book are given in Table 2.1. (Only the content of the Part is given if the title itself is inconveniently long.) References to the Code of Federal Regulations are normally given in shorthand form, "10CFR30.3," for example, meaning Title 10, Part 30, Section 3. The regulations of the other U.S. agencies of course are also published in the Code; 40CFR pertaining to the Environmental Protection Agency (EPA) and 49CFR to the Department of Transportation (DOT) as examples. The regulations are modified from time to time if new problems arise or an old rule turns out to inspire too much customer resistance. Notice of such proposed change is published in the *Federal Register* (a weekday daily newspaper put out by the U.S. Government Printing Office) in advance so that all interested parties may comment, usually by mail, but sometimes at public hearings called by the concerned agency as a means of being certain that all points of view are available before the regulation is officially adopted. Notice of the final adoption agains appears in the *Federal Register*. References to that publication are also given in shorthand, 37FR9207, for example, meaning Volume 37, page 9207. Both the *Register* and the Code of Regulations are issued by the U.S. Government Printing Office and are available through the Superintendent of Documents, Washington, D.C. 20402.

It will be seen from Table 2.1 that many of the 10CFR Parts are concerned with licensing, understandably enough, since licensing is the chief control mechanism available to the NRC. It should be noted that 10CFR30 lists quan-

Table 2.1 Nuclear Regulatory Commission Regulations

10CFR Part	Title or Content
1	(NRC Organization, General Information)
20	Standards for Protection against Radiation
30	(General rules for domestic licensing of by-product material)
31–33	(Different types of licenses for by-product material)
34	(Licensing and safety regulations for radiography)
35	Human Uses of By-product Material
40	Domestic Licensing of Source Material
51	Licensing and Regulatory Policy for Environmental Protection
70	Domestic Licensing of Special Nuclear Materials
71	(Packaging and transportation of radioactive materials)
150	(Division of authority between NRC and Agreement States)

Source: Title 10, U.S. Federal Code of Regulations.

tities for a large number of individual radioisotopes, below which level no license is required for use. Similarly, 10CFR40 exempts "unimportant quantities" of source materials where the U or Th content is below 0.05 wt%, unrefined ores, thorium in gas mantles, vacuum tubes, etc. (although there are quantity limits in some cases), and uranium metal used as shielding material in DOT-approved shipping casks. 10CFR70 does not list similar exemptions for special nuclear materials, but certain agencies (DOE, Department of Defense, etc.) and their contractors are not required to be licensed and persons holding general by-product licenses can utilize special nuclear materials in manufacturing calibration or reference standards.

The NRC has issued a number of "Regulatory Guides." Most of these deal with the arcane art of obtaining the various licenses needed in connection with proposed construction of reactors or other fuel cycle installations, but a few are of a more general nature. Complete lists are obtainable from the NRC.[14]

The Atomic Energy Act of 1954 authorized the AEC (and now the NRC) to enter into agreements with the individual States whereby the latter would assume regulatory control of by-product, source and less than critical amounts of special nuclear materials. 10CFR150 specifies the extent to which persons in Agreement States are exempt from NRC licensing and also identifies the activities over which the regulatory authority of the Commission continues. All of the States, Agreement or not, have regulatory powers over areas where the AEC-NRC has never had statutory authority; naturally occurring or accelerator-produced radioisotopes, for example. In addition, the States may carry out independent monitoring programs and participate in the licensing process through the individual State health departments. As things stand now (there are pressures to change) Agreement States regulate any commercially operated low-level radioactive-waste burial grounds within their borders. They formerly were also responsible for any abandoned uranium or thorium mill tailing (ore residues) piles in their state. Many of these piles originated with the World War II Manhattan Project and can be a problem because of their content of ^{226}Ra and/or other long-lived daughters of the U or Th decay chains, so the States have long been unhappy about their presence. In late 1978 Congress took "compassionate responsibility" and authorized the NRC to assume licensing authority for the abandoned piles as an extension of its existing control of actively operating mills. Congress also allotted funds to cover the major costs of cleaning up the orphan wastes.

In early 1979 there were 25 Agreement States with Michigan and Illinois negotiating contracts with the NRC.

A Federal Radiation Council was established in 1959 consisting of the Secretaries of Defense, Labor, Commerce, and Health, Education and Welfare (HEW). Its charge was ". . . to advise the President with respect to radiation matters directly or indirectly affecting health, including guidance to all Federal agencies in the formulation of radiation standards and in the development and execution of programs of cooperation with the states." The Council, modified in composition from time to time, functioned until EPA was formed by U.S.

Government Reorganization Plan No. 3 of 1970, which transferred the council's duties to the new agency. The responsibility of collating, analyzing and interpreting data on environmental radiation levels as collected by all of the monitoring networks in the country was simultaneously transferred to EPA from the Bureau of Radiological Health of the Food and Drug Administration (FDA), itself a part of the Public Health Service (PHS) of HEW. EPA also received the charge from the National Environmental Protection Act for reviewing the environmental aspects of all major Federal projects and of developing radiation criteria and standards for drinking water. PHS-FDA retains authority in relation to radiation in consumer products, in the healing arts and in occupational radiation protection.[15,16]

The very broad charge to the Radiation Council as it was transferred to EPA rather predictably raised some jurisdictional problems, one of which was the question of whether EPA or AEC should set standards for the individual facilities of the nuclear fuel cycle. There were several court hearings, with the problem eventually being arbitrated at the White House level. In essence, it was decided that AEC (NRC) controlled "within the fence" and EPA "outside the fence", i.e., NRC sets standards for effluents, emissions, wastes, etc., for the facilities it licenses, while EPA sets standards controlling the effect on the general environment of the fuel cycle operation as a whole.[16]

The roles of NRC, EPA, FDA and the States have all been briefly reviewed with many nuances necessarily omitted. DOE remains automonous in that it sets the standards within its own house. The Department of Transportation (DOT) controls transportation of radioactive materials across state lines, as does the Post Office for items sent through the mails. Other federal agencies can become involved in special cases, such as the Department of Commerce if application is made to build a reactor on coastal lands. The Bureau of Mines of the Department of the Interior is responsible for establishing radon-level standards to protect workers in uranium mines, and the Geological Survey is playing a role in selecting suitable sites for underground disposal of radioactive wastes. Projects involving radioactivity may thus require permits or other forms of appoval from a number of government organizations. In many cases elaborate environmental impact statements and safety analyses may be required. There is quite a contrast with the 1955 situation as reported by Dr. Taylor.

The complicated control system described seems to work fairly well, although frequently creating horrendous delays in getting things done. Periodic national meetings of agency representatives are held[17] to help with the communications problem.

2.6 THE RADIATIONS

Before discussing the radiation protection standards themselves, it would be well to delve a little way into the involved subject of dosimetry and dosimetric units, which in turn calls for a brief review of the types of radiation being considered and their reaction with matter. The review will be almost entirely descriptive and, for the purposes of this book, largely confined to radiations with energies of less than 10 MeV since particles or photons of greater energy are very rarely encountered in ordinary laboratory work with radioisotopes. Charged particles can of course be brought to very much higher energies in accelerators and such superenergetic projectiles will generate extremely high-energy x-rays and neutrons upon striking a target. The accelerator-associated radiation protection problem thus becomes a considerably expanded ballgame and no attempt will be made to discuss it here. The NCRP[17] and NBS[18] have both issued pertinent reports.

Health physicists classify radiations as either directly or indirectly ionizing, with "ionizing" covering either type or a mixture of both. The directly ionizing radiations are the charged particles such as electrons, alphas, positrons, protons, fission fragments, etc.; the indirectly ionizing group includes x- and gamma-rays and neutrons. (As a reminder, an electron originating in the nucleus is a beta-ray if the charge carried is negative, a positron if the charge is positive. A free electron originating outside the nucleus is simply that—an electron with a charge of -1. Similarly, if electromagnetic radiation originates in the nucleus, it is a gamma-ray. If the origin is in energy level changes in the electron cloud outside the nucleus, it is an x-ray.)

A charged particle moving at high speed is loaded with electrical-magnetic forces. When such a particle enters a target (anything but a high vacuum) it intrudes into an environment where similar forces are pervasive but at equilibrium. The particle coming into this nicely balanced electrical system is the proverbial bull in the china shop, wreaking its damage (or depositing energy in the right places in the case of beneficial applications) through Coulombic *interaction* with orbital electrons or nuclei of the atoms traversed. These interactions can cause ejection of an orbital electron from a target atom to form an ion pair (a free electron plus a positively charged atomic or molecular residue, i.e., ionization) or only cause excitation of the target element or molecule with subsequent photon emission of various wavelengths, frequently in the visible-light range, as the target electrons pushed into higher orbitals drop back into lower energy states in order to restore the status quo. Indirectly ionizing radiations, by contrast, lose their energy by *collisions* with orbital electrons and nuclei, the charged particles thus formed in the target material going on to produce ionization in the surroundings. The ionization produced by x- and gamma-rays and by neutrons is accordingly of a secondary nature.

Since charged particles lose their energy by mechanisms that do not require direct collisions, the energy-transfer process occurs rapidly as the particle moves into a target, the actual rate depending primarily on the initial energy, mass, and charge of the particle and on the nature of the target material. When these parameters are set, the heavier particles, such as alphas, will have a definite penetration depth in a given material, that is, a range within which the particle's energy will have been transferred completely. The situation with a beta particle is more complicated since it is much lighter and faster moving and may be diverted from the straight and narrow a number of times before disappearing. Nevertheless, a beta of a given energy will have a reasonably definite range although its overall path before reaching the end of the line may be tortuous and much longer than the range would indicate. Positrons are annihilated when they encounter a free electron (a "negatron") but there is some evidence[19] that the positron range is slightly greater than that of an equivalent electron.

The numerical value of a particle's range will be smaller as the mass and charge of the incoming particle increase. For particles of equal initial energy the range will decrease in the order: electron $\cong$ positron $>$ proton $>$ alpha $>$ fission fragment. Since this is the case, the amount of energy released per unit length of penetration [linear energy transfer (LET)] will be in the reverse order. These points are illustrated in Figures 2.1 and 2.2 taken from Ref. 20. Fission fragments have varying masses and perhaps charges so each will have a slightly different range. The two fragments produced in a binary fission event will share about 160 MeV of kinetic energy, so they are highly energetic. The range of an "average" fission product in aluminum is of the order of 3–4 mg/cm², equivalent to a penetration depth of about 15 μm.[21] With so much energy to dissipate in this distance the LET values are extremely high, and fission fragments, while very readily stopped in shielding terms, are extremely damaging in their brief careers.

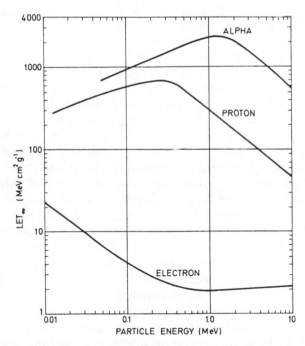

Figure 2.1   Range of charged particles in aluminum. [Dividing the ordinate values by the density of Al (2.70) and multiplying by 10 gives micrometers of penetration.] (From Ref. 20, Courtesy IAEA).

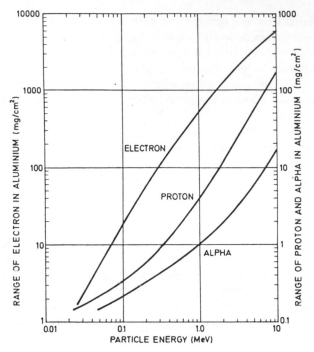

Figure 2.2 L_∞ values of charged particles in water. (Dividing the ordinate values by 10 gives keV/μm.) (From Ref. 20, Courtesy IAEA.)

The simplified discussion above implies that the rate of linear energy transfer is constant as a beam of charged particles enters a target. This is not the case as there can be scattering effects at the target surface along with other complications. In the case of an alpha particle, deceleration as it moves along its track means that there is more time towards the end for interaction with the surrounding atoms. This results in a "spike" in the LET near the end of the range. The LET then rapidly drops to zero as the particle accumulates neutralizing electrons and is converted to a neutral helium atom.

Even the most energetic radioisotope-produced alpha will barely penetrate the dead layer of the human skin, so alphas are a minor hazard from the point of view of external radiation. However, because of their high LET, alpha-emitters can be very damaging if taken within the body, particularly since most of them tend to deposit in the bones from which they are very slowly removed by metabolic processes. Beta particles are more penetrating than alphas, but even they—if not artificially accelerated—can be readily stopped by interposition of a thin sheet of shielding material. In this case, however, there are complications since energetic free electrons can generate x-rays (bremsstrahlung) as they move through matter. The implications of this phenomenon will be considered briefly in the section on shielding, as will the reactions of the indirectly ionizing radiations with matter.

Radiation impinging on or actually within a living form can produce many effects. DNA and RNA chains can be broken, cell enzymatic systems disrupted, and in the case of bone, an internal emitter can actually cause structural damage. As a very pertinent example, since living tissue is mostly water, radiation of H_2O has among its first effects the production of the hydrated electron, the ultimate in powerful chemical reducing agents, and the free OH radical, almost as potent an oxidant.[22,23] These, and the many other highly reactive chemical intermediates produced in water by irradiation, while mostly extremely short lived, can seriously impair normal cell mechanisms.

There are innumerable other parameters determining biological effects—length and level of exposure, mode of exposure (internal or external, tightly collimated beam or isotropic, etc.), different radiosensitivities of the various body organs, species and individual differences in resistance, mixtures of radiation types each with its own pattern of reaction, the body's repair mechanisms, and so on. The dosimetry field is thus very difficult and an area of continuing very active research. Several reference texts are listed in the Appendix.

2.7 RADIATION AND DOSIMETRY UNITS

The complexity referred to in the last paragraph has resulted in much rethinking over the last fifty years of appropriate units for use in radiation protection work, a process that still continues. Table 2.2 summarizes the units that will be discussed. The terminology, symbols, and quantity definitions are taken from the most recent ICRU publication[24] on radiation quantities and definitions. (ICRU Report 19, which also defines other units not reviewed here. Report 19 supercedes Report 10a, which was produced as NBS Handbook 84 in 1962, and Report 11 in 1968.) Each of the units will be defined and briefly explained if such seems necessary. The "special units" are the most familiar and will be discussed first before getting into the "SI" units question.

When the roentgen was adopted as the unit of exposure in 1928 the name was capitalized (Roentgen or Röntgen) and the symbol given in lower case (r). The usage in ICRU 19 is now reversed—roentgen and R. The 1928 definition applied only to x-rays but in 1937 was redefined to include gammas up to 3 MeV. The definition: "The roentgen shall be the quantity of x- or gamma-radiation such that the associated corpuscular emission per 0.001293 g of air [1 cm^3 at N.T.P.] produces, in air, ions carrying 1 esu of quantity of electricity of either sign." The definition was later recast to: ". . .the exposure of x- or γ-radiation such that the associated corpuscular emission per kilogram of air produces, in air, ions carrying 2.58×10^{-4} coulombs of energy of either sign." The two quantities are numerically the same and correspond to absorption of 86.9 ergs/g of air. The energy absorption in substances of different atomic number will change (about 98 ergs/g in tissue).[20,24]

There are several important points concerning the roentgen. First, the unit still applies only to x- or gamma-rays and cannot be used for charged particles or neutrons. Second, the roentgen is an *exposure* unit. A statement to the effect that a reading of x R/hr was measured at a certain point simply means that an

Table 2.2 ICRU Radiation and Dosimetry Units

Name	ICRU Symbol	SI Unit	Special Unit
Exposure	X^a	$C\,kg^{-1}$ (coulombs/kilogram)	R (roentgen)
Absorbed dose	D	$J\,kg^{-1}$ (joules/kilogram)	rad
Dose equivalent	H	—	rem
Quality factor	Q	—	—
Linear energy transfer	L_Δ	$J\,m^{-1}$ (joules/meter)	$keV\,\mu m^{-1}$ ($keV/\mu m$)
Kerma	K	$J\,kg^{-1}$ (joules/kilogram)	rad
Fluence		m^{-2} (per meter^2)	—
Flux density		$m^{-2}\,s^{-1}$ (meter^2/second)	—
Activity	A	s^{-1} (per second)	Ci (curie)

Source: Excerpts from Table 1 of ICRU Report 19.
^{a} $\dot{X}$ indicates delivery rate per unit of time. The same is true for $\dot{D}$ and $\dot{H}$.

individual standing at that point would be exposed to a defined quantity of photon radiation per unit of time. It says nothing concerning the amount of energy that would be absorbed into the individuals body or the resulting biological effects.

The rad (*r*adiation-*a*bsorbed-*d*ose) was introduced by the ICRU in 1953[25] as a unit that would depend only on the amount of energy absorbed per unit mass of material, irrespective of the nature of the material or of the energy or type of the radiation. It is defined as the amount of energy imparted to matter by ionizing particles per unit mass of irradiated material at the point of interest. It is thus the *absorbed dose*, with 1 rad equal to the absorption of 100 ergs/g.

While the absorbed dose measures the amount of energy imparted to an irradiated material, it still does not define the biological effects of that absorption, the quantity of primary interest to the radiobiologist or health physicist. These effects can vary with the type of radiation, as previously discussed, and with the energy-release mechanisms within the body. A new unit seemed to be needed and brought about the concept of "dose equivalent." The term originated from discussions between the ICRU and ICRP in 1962 and the most current definition has been given in ICRU 19: "The dose equivalent, H, is the product of D, Q, and N at the point of interest in tissue, where D is the absorbed dose, Q is the quality factor and N is the product of any other modifying factors

$$H = DQN."$$

The equation, which will be discussed further below, adjusts the absorbed dose for the fact that different types and energies of radiation release that energy at different rates as they pass through tissue, and also leaves a little elbow room for handling other possible variables. There is evidence that some of the organs of the body are more radiosensitive than others and consideration has been given to including this variable as a separate correction factor in the equation. The idea has not been adopted, however, since the background information needed for such an application is still too limited.

Use of the dose equivalent apparently still generates some confusion[26,27] and the ICRU has issued two subsequent clarifications, "Dose Equivalent, Supplement to ICRU Report 19"[28] in 1973 and "Conceptual Basis for the Determination of Dose Equivalent"[29] in 1976.

The dose equivalent is expressed in terms of rems (*roentgen-equivalent-man* or *H*) and was originally defined as ". . . that dose of any ionizing radiation which, when delivered to man or mammal, is biologically equal to the dose of one rad of x- or gamma-radiation." The rad to rem conversion at that time was expressed as

$$\text{dose in rems} = \text{dose in rads} \times \text{RBE},$$

where RBE is the relative biological effectiveness, defined as the quantity of 250-keV x-rays needed to produce a given biological effect divided by the quantity of the radiation of interest producing the same effect. Use of the RBE factor in both radiobiology research and in radiation protection however generated certain problems, so RBE is now reserved for the former area and Q (the quality factor, previously seen as QF) for the latter. Q depends only on the rate of linear energy transfer, a term which has been previously discussed, and which will be seen in the literature as LET, L_δ, L_∞, etc., depending on the restrictions put on the energy range of the radiation considered in the definition. The other multiplying factor, N, appears to be in the catchall category— if there are peculiar conditions present, such as uneven distribution of radiation in an internal organ, the dose in rads can be appropriately adjusted through N to obtain the dose in rems. The factor is rarely used in everyday work.

The kerma is a specialized dosimetric measure of radiation fields for uncharged particles such as neutrons or x-rays. It is equal to the sum of the kinetic energies of the charged particles generated per unit mass with which the uncharged particles react. The kerma unit is again the rad. Under some conditions the kerma will equal the absorbed dose, under others there will be some differences as discussed at length in ICRU Report 26.[30]

From the foregoing discussion it will be seen that there has been a continuing evolution of dosimetric concepts and units, and it is still underway. The problem of course is that a radiation field can be reasonably well defined and measured outside the body, but this is the only information available for estimating the effect on different organs at different depths in the body, and the situation becomes even more complicated if the emitter is actually within the organism. For incident monoenergetic radiation of a given type, the dose equivalent tends first to increase with depth, then to reach a maximum and finally to decrease (see the alpha-ray discussion, Sec. 2.6). The increase may be due to buildup of charged particles from an incident uncharged particle beam, to scattering of the incident radiation near the entrance portion of the body or to production of secondary radiations of other types. The final decrease of the dose equivalent results from attenuation of the primary beam. The position of the maximum dose equivalent in the body depends upon the energy and type of radiation.[28,31]

In order to deal with the practical problem of locating the point of maximum dose equivalent, the ICRU has defined[24,29] two additional terms, the absorbed dose index (D_1) and the dose equivalent index (H_1). These indices are based on the location of the maxima of the two dose units in a 30-cm-diam sphere of density 1 (a "phantom"), chosen as an approximation of the trunk of the human body. The rationale behind these indices will not be attempted here, but is given at some length in Ref. 29. The practical intent of the indices is to provide a better specification of the relationship between measurements made in the volume of interest with no receptor present and the effects on a receptor (an individual) entering that volume. Among the applications would be specification of exemption limits for sealed radioactive sources, packages accepted for general transportation, etc.

The remaining definitions in Table 2.2 are quite straighforward. The fluence is the number of particles entering a sphere of a stated cross section and the flux density expresses the fluence rate [as for $n/cm^2 \cdot$ sec (neutrons per square centimeter per second) in reactor work]. The curie (now lower case c and symbol Ci rather than the old C) is the unit of radioactivity quantity and was developed simultaneously with the concept of the roentgen. The curie was originally defined as the disintegration rate from 1 g of radium. Direct and exact measurement of this quantity was difficult, so the unit was redefined as the disintegration rate of the amount of radon in equilibrium with 1 g of radium, then changed in 1930 to the disintegration rate of *any* daughter in equilibrium. This number came out to be very close to 3.7×10^{10} disintegrations (spontaneous nuclear transformations) per second and this is the quantity that is now officially accepted—at least temporarily. Further change is probable because of the presence of SI units. These have been around since 1960[32] but only now appear to be rapidly coming over the horizon in radiation terms.

The International System of Units (abbreviated SI) was developed by the CGPM (Conférence Générale des Poids et Mesures, or General Conference on Weights and Measures) as a coherent international measurement system to cover all of science and engineering and based on seven base units (meter, kilogram, second, ampere, kelvin, mole and candela) and two supplementary units (radian and steradian). All other units are to be derived from these basics by simple multiplication, division or the use of exponents.[32] Thus the unit of force becomes the newton (N), or kg $\cdot$ m/sec^2 and that of energy, the joule or N$\cdot$m. The SI is an update and expansion of the metric system.

The ICRU has naturally considered just how radiological units might fit into the SI framework for almost twenty years. Some of the difficulties are obvious, but matters now appear to be going into high gear. In 1973 the Commission published articles[33] asking for worldwide comment on the changes that would be required. The replies were considered at the 1974 meeting of the ICRU, following which a letter was submitted to the CGPM asking for adoption of the "gray" as the SI unit of absorbed dose and related quanities (expressed as joules/ kilogram) and the "becquerel" [one disintegration (dis) per second] to replace the curie as the quantity unit for radioactive transformation. The CGPM approved both new names and the radiological community was so informed by a letter sent by the ICRU to a number of professional journals.[34] In the letter, the Commission pointed out that they had asked that the roentgen and the curie be temporarily retained as "special units," but that they should be gradually phased out to be replaced by the gray and becquerel over the following ten years.

The proposed changes have received very mixed reviews by users, although the new units had already been officially accepted by the Germans and Belgians.[35] Nevertheless, there was not universal dancing in the streets, the objections being primarily that the becquerel at 1 dis/sec would be inconveniently small and the gray (equivalent to 100 rads) similarly awkwardly large. There is also the not inconsiderable fact that the roentgen and the curie have become throughly embedded in the literature and in user thinking over the last fifty years. This latter problem is of course similar to the one that the United States confronts as it considers conversion to the metric system.

Table 2.3 (from Ref. 35) summarizes the proposed unit changes. The rem will also have to change and the name "sievert" (Sv) has been proposed and apparently already adopted by the ICRP. The unit would be related to the rem in the same manner as the gray to the rad, that is 1 Sv would be 1 J/kg and equal to 100 rems. The absorbed dose and dose equivalent indices will also have to change.

The quality factor is assumed to be 1 for x-rays and gammas so the absorbed dose is equal to the dose equivalent for these photons. Over the 100 keV to several MeV range the exposure in roentgens for these radiations is also

Table 2.3 Proposed SI Units and Their Present Equivalents

Quantity	Derived SI Unit	SI Name and Symbol	Present Equivalent
Activity	per second (sec^{-1})	becquerel (Bq)	2.703×10^{-11} Ci
Absorbed dose	joules/kilogram (J·kg^{-1})	gray (Gy)	100 rad
Absorbed dose rate	watts/kilogram (W·kg^{-1} = J·kg^{-1} sec^{-1})	Gy/sec	100 rad/sec
Exposure	coulombs/kilogram (C·kg^{-1})	—	3.876×10^3 R
Exposure rate	amperes/kilogram (A·kg^{-1} = C·kg^{-1}·sec^{-1})	—	3.876×10^3 R/sec

Source: Modified from Ref. 35.

Table 2.4   Relationship between Linear Energy
Transfer and the Quality Factor

L_∞ in Water (keV/μm)	Quality Factor (Q)
3.5 or less	1
7.0	2
23	5
53	10
175	20

Source: Reference 24.

roughly equal to the absorbed dose in rads (Ref. 36, p. 10). For electromagnetic radiation in this range it can therefore be assumed for most practical purposes that roentgens = rads = rems.

The quality factor for other radiations depends on the rate of linear energy transfer. The relation of L_∞ to Q is given in Table 2.4. For betas, Figure 2.2 shows that L_∞ is always below 3.5 keV/μm (at least up to 10 MeV) so the quality factor is again 1 and rads = rems. Q values for alphas and protons in the 10-MeV range can be approximated by comparing Table 2.4 and Figure 2.2. Quality factors for neutrons up to 10 MeV are given in Table 2.5. In

Table 2.5   Conversion and Quality Factors for Neutrons[a]

Neutron Energy (MeV)	Conversion Factor [(n/cm$^2 \cdot$ sec)/(mrem/hr)]	Effective Quality Factor (Q)
2.5×10^{-8} (thermal)	260	2.3
1×10^{-7}	240	2
1×10^{-6}	220	2
1×10^{-5}	230	2
1×10^{-4}	240	2
0.001	270	2
0.01	280	2
0.1	48	7.4
0.5	14	11
1	8.5	10.6
2	7.0	9.3
5	6.8	7.8
10	6.8	6.8
20	6.5	6.0

Source: Table 4, Appendix 6, Ref. 37.

[a] Both factors are calculated at the maximum of the depth−dose equivalent curve.

practical radiation work the exact energy of the radiation of interest may be uncertain or there may be a mixture of energies. In such cases it is usually adequate to assume Q values of 1 for x-rays, gammas and betas; 3 for thermal neutrons; 10 for fast neutrons, protons and radioisotope-produced alpha radiation; and 20 for heavy recoil nuclei.[38,39]

2.8 PROTECTION STANDARDS

Very little has been said to now concerning the effects of radiation exposure upon an individual, the key concern of the various commissions and regulatory agencies that have been discussed. The problem here is that while overt clinical responses to large doses are relatively easy to define, there is still considerable uncertainty in relation to the more subtle effects such as cancer activation, possible lifeshortening, accelerated aging or genetic damage affecting future generations, particularly in connection with low radiation doses delivered over a sustained period of time.

For acute (very brief period) x-ray or gamma exposure, IAEA gives[20] the following summary:

Acute Dose Level (rads)	Probable Effect
0–25	No obvious injury
25–50	Possible blood changes, but no obvious injury
50–100	Blood cell damage, some injury, no disability
100–200	Injury, possible disability
200–400	Injury and disability certain, death possible
400–500	50% fatal within thirty days
600 or more	Probably fatal

Information of this rather qualitative nature was essentially the total available to the ICRP and NCRP initially, but as more data accumulated, particularly on possible genetic damage, it became obvious that a very conservative approach should be taken in establishing control standards. The ICRP now bases its recommendations on the year, with limits as to the amount of exposure in any one quarter.[41] The recommendations also take into account the fact that different parts of the body have different susceptibilitites to radiation damage. Table 2.6 presents the current standards under two headings, the first applying to individuals handling radioactivity in the course of their daily activities ("radiation workers" such as doctors, nurses, fuel cycle employees, researchers, etc.) and the second to members of the general public. The maximum permissible doses (MPD's) for the two groups differ by a factor of 10. This is not because radiation workers are necessarily considered to be more

Table 2.6   ICRP Dose Limits for Individuals[a]

Organ or Tissue	Maximum Permissible Doses (rems/yr)	
	Radiation Workers	General Public
Gonads, red bone-marrow	5	0.5
Skin, bone, thyroid	30	3
Hands and forearms, feet and ankles	75	7.5
Other single organs	15	1.5

Source: Modified from Ref. 31, Appendix.

[a] References 41 or 31 should be consulted for qualifying statements.

expendable, but because the general population includes infants, children, and fetuses whose susceptibility to radiation is greater than that of adults.[42] There is currently considerable pressure to reduce the permissible exposure limits even further, partly based on the findings of the 1979 BEIR report.[8]

In the case of radiation workers, a half-year's dose can be taken in a single quarter if the 5 rems/yr overall limit is not exceeded. The quarterly dose can be acquired on a regular basis, if necessary, but under no circumstances should the total dose accumulation for the years since the individual's eighteenth birthday be allowed to exceed 5 $(N - 18)$, where N is the current age. The exposure to pregnant women should not exceed 0.5 rem during the entire gestation period.[43] The dose limit for the thyroids of persons under sixteen years of age is 1.5 rems/yr.[31,37] (In 10CFR20.104 the NRC uses the same limit for minors as for members of the general public, that is, at $\frac{1}{10}$ the level of other radiation workers.) These limitations are essentially the same as those adopted by the NRC in 10CFR20 and also appear in ERDA Manual chapter 0524,[44] that former agency's own internal code of regulations, apparently still adhered to by the DOE.

The next step in the standards setting procedure is to establish rules such that individuals, again subdivided into radiation workers and members of the general public, will not be exposed to radiation levels that would result in dosages higher than the limits given above. For a purely external hazard, such as a clad metallic ^{60}Co source, the calculations needed to avoid overexposure, while not always simple, are relatively straightforward since the exposure field can usually be readily measured. The problem becomes much more complicated for radioactivity taken into the body. Such introduction is most likely to be by inhalation of radioactive gas or particles from the atmosphere or by ingestion of soluble or suspended material in water. (Other less probable routes would be through consumption of contaminated food or direct transfer through the skin, in a few rare cases, or through breaks and wounds.) The radioisotope may be in an insoluble form that will pass through the digestive system or removed in the lungs, or may be soluble and enter rapidly into the bloodstream. Once in the body each active element is metabolized differently depending on its chemistry, and may tend to concentrate in a particular organ or organs, the notable example being the movement of iodine to the thyroid. Some elements are excreted from the body more rapidly than others, leading to the concept of "biological half-life" or the length of time needed for the body to eliminate one half of its content of the radioisotope in question. For the isotopes of some elements, these half-lives are far in excess of a human lifetime. A radioactive species having both a long physical (decay) and a long biological half-life is obviously an undesirable passenger to have on board.

The basic approach taken by those specifying standards has been to consider the variables enumerated above in connection with the individual isotopes and then to set concentration limits in air and water for each (over and above the natural background radiation) such that continuous exposure (40 h/week for radiation workers and 168 h/week for the public) would not result in greater dosages than the limit established for the particular organ of the body most susceptible to damage by the radiations of that particular isotope. The calculations made by the ICRP[45] in deriving the air and water concentration limits are based on the assumption that the 40-hr/week or 168-hr/week exposures will be continuous over a 50-year period and that in that time the concentration in the critical organ would have built up to the estimated potentially dangerous level if the exposure at the limit was uniform throughout.

Thus for a given isotope the recommended air and water concentration tables may include the specification of up to ten sensitive organs, the microcuries of the activity representing the maximum permissible burden for the total body and then the maximum permissible concentrations (MPC's) of the isotope in the air breathed in by an individual and in the water that he might drink. Each of these items is usually given for the isotope in both its soluble and insoluble forms. The whole pattern is repeated twice, once for radiation workers and once for members of the general public. There are also special rules applying to mixtures of isotopes, to emitters of unknown nature and to the situation where the danger of heavy-metal poisoning may actually be greater than that from the radioactivity, such as with some of the isotopes of uranium. Exposure to a radioactive gas, where there will be both inhalation and external exposure, is treated as if the individual were submerged in a half-hemisphere cloud of the gas of semi-infinite extent. The critical organ in this case is usually the whole body or the skin.

Since the MPC's are concerned with internal radiation there is no point in evaluating extremely short-lived or extremely rare activities, although some of the former can be significant sources of external radiation. MPC's for all of the known isotopes accordingly have not been calculated, but the recommendation lists are still quite long. The basic tables are found in ICRP 2,[45] later modified in the case of ^{90}Sr and ^{238}U and with new transuranium isotopes added in ICRP 6,[42] and further assessments of the mode of intake into the body (one-time versus recurrent exposure, for example) made in ICRP 10[46] and ICRP 10a.[47]

The NRC,[48] DOE[44] and IAEA[38] use the ICRP numbers verbatim for radiation workers, and the last named of those agencies restricts itself entirely to that part of the tables. NRC and DOE do however differ from ICRP in the tabulated limitations on exposure levels to the general public, establishing limits more conservative in most cases by factors of 2 to 10. (The reason for this is not clear, but presumably associated with a recommendation in Paragraph 65 of ICRP 2 that isotopes tending to concentrate in a particular organ be given a MPC $\frac{1}{30}$ rather than $\frac{1}{10}$ of the occupational value. Table 6 in NCRP 39 similarly indicates a reduction of 30 for genetic and somatic dose limits to the public.) Neither NRC or DOE present any body burden or critical organ data in their tables, but always choose the most stringent of the ICRP levels before applying their own further reductions. The IAEA uses the terms "derived water concentration" and "derived air concentration" rather than MPC.

It is interesting to note that the bulk of the ICRP recommendations were established in 1959 and that the approach used was sufficiently conservative so as to require very few modifications since. The plutonium in water MPC is however now being questioned since there is recent evidence[49] that the rate of gastrointestinal absorption of Pu in mice is independent of the element's oxidation state and may be as much as two orders of magnitude higher than previously assumed.

2.9 "ALARA"

The standards that have been discussed above are all expressed in concrete numbers—limiting dose per year, limiting concentrations in air and water, etc. Through constant usage, numbers of this type can become taken for granted, and it is then easy to assume that no significant injury will result if the limits are not exceeded. Such is quite possibly true, but no one can be absolutely sure in light of present knowledge. Since the question of the linear relationship between low-level radiation exposure and biological damage is still uncertain and hotly debated, the ICRP, while feeling that its standards have been established on a very conservative basis, has nevertheless consistently taken the reasonable stand that all unnecessary exposure should be avoided, even if it is within its own established limits. In recent years the Commission has accordingly been emphasizing ALARA—"As Low as Reasonably Achievable, economic and social considerations being taken into account"—a term much heard in health physics circles these days. Such a qualitative concept as contrasted to positive numerical limits has caused some confusion so the ICRP has published a special report[50] giving the philosophical background for their current emphasis on ALARA.

Numerical limits, in addition to producing a perhaps false sense of security, can also generate problems in the other direction if tunnel-view practitioners apply them too rigidly in all situations. The ICRP recognizes this real life contingency, and ALARA has as one of its aims the provision of more latitude to the knowledgable professional in the field for making reasonable decisions in particular situations, in effect saying that common sense is still a very valuable commodity when it comes to applying any set of rules.

ALARA thus has some definite cost-benefit overtones for radiation protection, and ICRP 22 devotes appendix material to this aspect. This attention to the practicalities certainly is partly due to the pressures imposed by the profusion of new regulations calling for elaborate environmental impact statements and safety analyses for even relatively trivial nuclear industry activities. The authors of such reports must quantify their statements as much as possible so that reasonable comparisons can be made between the potential hazards and the costs and the benefits to be gained by alternative courses of action. Such calculations must of course be based on existing radiation protection standards.

These efforts to quantify the imponderable future must necessarily be statistical in nature and have quite predictably introduced new terms into the language such as "population at risk" and "man-rem." These can probably be best explained by examples. EPA[1] estimates that the average whole body dose in the United States due to natural radiation is 130 mrem per person and that the U.S. population will be 237 million in 1980. Multiplying these two numbers together gives 30.8 million man-rem. In this case, everyone in the country is part of the population at risk. On the other hand, a mishap in a reactor in an isolated location might be projected as affecting 500 people in the plant and surrounding area at most, the population at risk for that situation. If it is estimated that the average dose received by each individual is 200 mrem, the man-rem total then becomes 100. Assume that there are two preventative measures that can be taken at the plant to reduce the effects of that type of mishap, the first costing $20,000 and predicted to reduce the man-rem total to 50; the second costing $180,000, but expected to cut the man-rem total to 10 if an accident occurs. The first method thus would cost $400 per man-rem reduction, the second $2000. This kind of estimating appearing in a safety analysis provides the licensing agency with background as to whether the more or less expensive alternate preventative measure should be mandated. There are a number of suggestions as to the magnitude of the trade-off figure in the literature, a recent one being $1000 per man-rem,[51] i.e., at a higher figure the loss in manpower and resources could not be justified by the benefits gained. Such an approach is one method of defining the "reasonable" part of ALARA.

The 1977 ICRP report[52] (Publication 26) deals with the question of cost-benefit calculations at some length.

REFERENCES

1   A. W. Klement, Jr., C. R. Miller, R. P. Minx and Bernard Schleien, "Estimates of Ionizing Radiation Doses in the United States, 1960-2000." USEPA Report ORP/CSD 72-1 (August 1972).

2   K. Z. Morgan, "History of Radiation Protection." *Mater. Eval.*, 29, 19A *et seq* (March 1971).

3 L. S. Taylor, "History of the International Commission on Radiological Protection (ICRP)." *Health Phys.*, 1, 97-104 (1958).

4 L. S. Taylor, "History of the International Commission on Radiological Units and Measurements (ICRU)." *Health Phys.*, 1, 306-314 (1958).

5 L. S. Taylor, "The Achievement of Radiation Protection by Legislative and Other Means." Proceedings of the *Int. Conference on Peaceful Uses of Atomic Energy, Geneva, 1955*, 13, 15-21 (1956), A/CONF 8/P319. United Nations, New York.

6 L. S. Taylor, "Brief History of the National Committee on Radiation Protection and Measurements (NCRP) Covering the Period 1929-1946." *Health Phys.*, 1, 3-10 (1958).

7 National Council on Radiation Protection and Measurements, "Radiation Protection Guidelines for 0.1-100 MeV Particle Accelerator Facilities." NCRP Report No. 51 (1977). NCRP, 7910 Woodmount Ave., Washington, D.C. 20014.

8 Advisory Committee on the Biological Effects of Ionizing Radiations, "The Effect on Populations of Exposure to Low Doses of Ionizing Radiation." National Academy of Sciences-National Research Council, Washington, D.C. (1979). Also see E. Marshall, "NAS Study on Radiation Takes the Middle Road." *Science*, 204, 711-714 (1979).

9 United Nations Scientific Committee on the Effects of Atomic Radiation, "Ionizing Radiation, Levels and Effects." Official Records, U.N. General Assembly, 27th Session, Suppl. No. 25, U.N. Document A/8725. In two volumes. United Nations, New York (1972).

10 I. G. Wilms and C. E. Moss, "Bookshelf on Radiological Health." *J. Public Health*, 65, 1231-1237 (1975).

11 10CFR30.4(d).

12 10CFR40.

13 10CFR70.

14 U.S. Nuclear Regulatory Commission (NRC), "Regulatory Guides." NRC, 1717 H Street, N. W., Washington, D. C. 20555.

15 W. D. Rowe, F. L. Galpin and H. T. Paterson, Jr., "EPA Environmental Assessment Program." *Nucl. Safety*, 16, 667-681 (1975).

16 W. D. Rowe, "EPA's Role in Standards Setting." In R. A. Karam, Ed., *Energy and the Environment Cost Benefit Analysis*, Pergamon, Oxford (1976), pp. 655-679.

17 Eighth Annual Conference on Radiation Control, Springfield, IL, May 2-7, 1976, "Radiation Benefits and Risks, Issues and Options." Report FDA-77-802 (April 1977).

18 U.S. National Bureau of Standards (NBS), "Radiological Safety in the Design and Operation of Accelerators." NBS Handbook 107 (1970). NBS, Gaithersburg, MD; mailing address, Washington, D.C. 20234.

19 P. S. Takhar, "Direct Comparison of the Penetration of Solids by Positrons and Electrons." *Phys. Lett.*, 23 219-222 (1966).

20 IAEA (Chap. 1, Ref. 2), "Radiation Protection Procedures." IAEA Safety Series No. 38 (1973).

21 S. Kahn and V. Forgue, "Range-Energy Relation and Energy Loss of Fission Fragments in Solids." *Phys. Rev.* 163, 290-303 (1967).

22 E. J. Hart and J. W. Boag, "Absorption Spectrum of the Hydrated Electron in Water and Aqueous Solutions." *J. Am. Chem. Soc.*, 84, 4090-4095 (1962).

23 M. S. Matheson and L. M. Dorfman, *Pulse Radiolysis*. M.I.T., Press, Cambridge, MA (1969).

24 International Commission on Radiation Units and Measurements (ICRU), "Radiation Quantities and Units." ICRU Report 19 (1971), ICRU, 7910 Woodmount Ave., Washington, D.C. 20014.

25 Anon, "Recommendations of the International Commission for Units." *Radiology*, 62, 106-109 (1954).

26   Alessandro Rindi, R. H. Thomas and Kurt Lidén, "Dose Equivalent and Absorbed Dose." *Phys. Med. Biol.*, 19, 738-739 (1974).

27   B. M. Wheatley and T. E. Burlin, "Modifying Factors and Dose Equivalent." *Brit. J. Radiol.* 45, 785-786 (1972)

28   ICRU, (Ref. 24 above) "Dose Equivalent, Supplement to Report 19," (1973).

29   ICRU, "Conceptual Basis for the Determination of Dose Equivalent." ICRU Report 25 (1976).

30   ICRU, "Neutron Dosimetry for Biology and Medicine." ICRU Report 26 (1977).

31   International Commission on Radiation Protection (ICRP), Committee 3, "Protection Against Ionizing Radiation from External Sources." ICRP Publication 15 (1970); Suppl., ICRP Publication 21 (1979), Pergamon, Oxford.

32   American Society for Testing and Materials (ASTM), "Standard for Metric Practice." Report ANSI/ASTM E-380-76, IEEE Stnd. 268-1976, ASTM, 1916 Race St., Philadelphia, PA 19103.

33   K. Lidén, "SI Units in Radiology and Radiation Measurements." *Radiochim. Acta*, 19, 151-152 (1973); "Special Radiation and/or SI Units?." *ibid.* 19, 210 (1973).

34   H. O. Wyckoff, A. Allisy and K. Lidén, "The New Special Names of SI Units in the Field of Ionizing Radiations." *Brit. J. Radiol.*, 49, 466-467 (1976); *Radiochim. Acta*, 22, 95-96 (1975); *Health Phys.*, 30, 417-418 (1976); *Radiology*, 118, 233-234 (1976).

35   W. A. Jennings, "SI Units in Radiation Measurement." *Brit. J. Radiol.*, 45, 784-785 (1972).

36   Herbert Goldstein, "The Attenuation of Gamma Rays and Neutrons in Reactor Shields." USAEC Div. of Reactor Development (1957); *Fundamental Aspects of Reactor Shielding*, Addison-Wesley, Reading, Mass. (1959).

37   ICRP (Ref. 31 above), Committee 3, "Data for Protection against Ionizing Radiation from External Sources: Supplement to ICRP Publication 15." ICRP Publication 21, Pergamon, Oxford (1973).

38   G. J. Appleton and P. N. Krishnamoorthy, "Safe Handling of Radioisotopes, Health Physics Addendum." IAEA (Chap. 1, Ref. 2) Safety Series No. 2 (1960).

39   IAEA (Chap. 1, Ref. 2), "Safe Handling of Radionuclides, 1973 Edition." IAEA Safety Series No. 1 (1973).

40   U.S. Department of Health, Education and Welfare (HEW), Public Health Service (Chap. 1, Ref. 7), "Radiological Health Handbook, Revised Edition," (1970).

41   ICRP (Ref. 31 above), Recommendations Adopted September 17, 1965." ICRP Publication 9, Pergamon, Oxford (1966).

42   ICRP, "Recommendations as Amended 1959 and Revised 1962." ICRP Publication 6, Pergamon, Oxford (1964).

43   NCRP (Ref. 7 above), "Basic Radiation Protection Criteria." NCRP Report No, 39 (1971).

44   U.S. Energy Research and Development Administration (ERDA), ERDA Manual (ERDAM), Part 0500, Health and Safety; Chapter 0524, "Standards for Radiation Protection." March 30, 1977.

45   ICRP (Ref. 31 above), Committee 2, "Recommendations on Permissible Doses for Internal Radiation (1959)." ICRP Publication 2, Pergamon, Oxford (1960). Also reproduced in *Health Phys.*, 3, 1-380 (1960).

46   ICRP, Committee 4, "Evaluation of Radiation Doses to Body Tissues from Internal Exposure Due to Occupational Exposure." ICRP Publication 10, Pergamon, Oxford (1968).

47   ICRP, Committee 4, "The Assessment of Internal Contamination Resulting from Recurrent or Prolonged Uptakes." ICRP Publication 10a, Pergamon, Oxford (1971).

48   10CFR20, Appendix B.

49 R. P. Larsen, R. D. Oldham, M. H. Bhattacharyya, E. S. Moretti and D. J. Austin, "Gastro-intestinal Absorption of Plutonium." *Radiat. Res.* (to be published.)

50 ICRP (Ref. 31 above), Committee 4, "Implications of Commission Recommendations that Doses be Kept as Low as Reasonably Achievable." ICRP Publication 22, Pergamon, Oxford (1973).

51 NRC (Ref. 14 above), "Cost-Benefit Analysis for Rad-Waste Systems for Light Water Cooled Power Reactors." NRC Regulatory Guide 1.110 (1976).

52 ICRP (Ref. 31 above), "Recommendations of the International Commission on Radiological Protection." ICRP Publication 26, Pergamon, Oxford (1977).

chapter 3

Radioisotope Laboratory Design

Present-day technology has made possible essentially routine handling of numerous dangerous substances—deadly viruses and bacteria, mutagenic or carcinogenic chemicals, explosives, recombinant DNA—in addition to radioactivity. The one aspect common to these varied activities if they are to be carried out safely is the availability of properly designed, constructed, and equipped facilities. These are not inexpensive to build, but very necessary if life and property are to be protected. Adequate funds and skilled engineering help were made available immediately following World War II to get the fledgling nuclear effort on its way, a fact that without question has had much to do with the unusually good safety record[1] established by the AEC, its successors and their contractors since 1946.

A rather arbitrary decision has been made in this book to treat radioactivity laboratories and shielded facilities in separate chapters in order to reduce the subheadings to a reasonable number. Some trade jargon will creep in. A "hot" or "dirty" area is one where radioactive contamination is actually or potentially present, while "cold" and "clean" of course have the opposite meaning. "Warm" or "suspect" is sometimes applied to buffer areas between the hot and cold parts of a facility. The term "hot laboratory" has come to imply the presence of some sort of shielded enclosure, rather than a laboratory as such.

3.1 HISTORICAL

Experience in handling radioactivity in any quantity was very limited when work began on the development of atomic energy in the early 1940s. Up to that time experimentation with radioactive materials had usually been undertaken in facilities very little different from those used for other chemical operations. World War II pressures and limitations permitted only modest change in this approach, but after peace was declared both time and experience were available to properly evaluate the special requirements for working with active nuclides. That the principles established and decisions made were generally satisfactory is demonstrated by the fact that relatively little has appeared on overall radiochemistry laboratory design philosophy since that time.

The material quoted in the above paragraph (slightly modified) was taken from the precursor chapter[2] to the present book. That chapter appeared in 1963. A review of the literature since that date still reveals no major revolution in radioisotope laboratory design, although there have of course been improvements in the construction materials, equipment, and control instrumentation available. The general design philosophy itself has withstood the wear and tear of some thirty years surprisingly well, and the basics will probably stand up indefinitely. Changes in application may however be in the offing. Radioactivity handling facilities as many of them now exist can be profligate users of energy, a fact that is already generating considerable rethinking of engineering approaches.

Among the older references on radioisotope laboratory design are a series of papers appearing in *Industrial and Engineering Chemistry* in 1949[3] and a summary report[4] of a conference held about the same time. Two books on general laboratory design have been published under the auspices of the National Academy of Sciences–National Research Council.[5,6] Each contains articles pertaining to radioisotope facilities. Volume 1 of Fitzgerald's book[7] contains a section on the design of nuclear facilities. Series of brief articles on general laboratory design have appeared both in *Analytical Chemistry*[8] and in the *Journal of Chemical Education.*[9] Interesting short papers[10] on laboratory design have also appeared in various semitrade periodicals.

3.2 DEFINING THE RADIOISOTOPE LABORATORY

There have naturally been official attempts to spell out the differences between the radioisotope laboratory and its more usual counterpart. The IAEA[11,12] first categorizes the various radionuclides on a relative hazard basis: very high radioactivity, high toxicity, moderate toxicity and low toxicity. For each of these four groups the Agency then specifies the type of laboratory required (A, B or C) for working with a prescribed quantity of the activity, given as a limit or a range. These specifications are given in Table 3.1. (The footnote of the table as given in IAEA SS No. 1[11] is partially repeated here: "Type C is a good quality chemical laboratory. Type B is a specially designed radioisotope laboratory. Type A is a specially designed laboratory for handling large quantities of highly radioactive materials. In the case of a conventional modern chemical laboratory with adequate ventilation and fume hoods, as well as polished, easily cleaned, nonabsorbing surfaces, etc., it would be possible to increase the upper limits of activity for Type C laboratories towards the limits for Type B laboratories for toxicity groups 3 and 4.")

Table 3.1 IAEA Radioactivity Level Recommendations[a]

Radiotoxicity Group	Laboratory Type Required		
	Type C	Type B	Type A
1. Very high	$<10\,\mu Ci$	$10\,\mu Ci - 10\,mCi$	$>10\,mCi$
2. High	$<100\,\mu Ci$	$100\,\mu Ci - 100\,mCi$	$>100\,mCi$
3. Moderate	$<1\,mCi$	$1\,mCi - 1\,Ci$	$>1\;\;Ci$
4. Low	$<10\,mCi$	$10\,mCi - 10\,Ci$	$>10\;\;Ci$

Source: Condensed from Table II, IAEA Safety Series No. 1 (Ref. 11).

[a] IAEA Safety Series Nos. 1 and 38 were both published in 1973, but the recommendations in the latter are appreciably less conservative than those shown above.

IAEA SS No. 1 then goes on to describe the various laboratory types, although in quite general terms. The Agency also supplies modifying factors that can be applied to the limits shown in Table 3.1, depending upon the type of operation being undertaken. These modifying factors are given in Table 3.2.

The American Institute of Chemical Engineers issued a design guide for Type-B laboratories[13] through the American National Standards Institute (ANSI). That earlier standard has been followed by ISO/R1710,[14] issued by ANSI acting as Secretariat for the International Organization for Standardization. Both of these documents are again somewhat general in nature, being primarily recommendations and reminders of potential problem areas.

The AEC had its own internal code of regulations known as the AEC manual (AECM) as indicated in Chapter 2. There basically was only a title change when AEC was eliminated, and the document became ERDAM (Chap. 2, Ref. 44). Now that ERDA has also disappeared and DOE is in charge the situation is not very clear. The agency has apparently voided large portions of ERDAM but has stated that the manual should still be used "for guidance" by its contractors. References in this volume will of necessity be to ERDAM.

Table 3.2 Conditions Modifying Table 3.1 Levels

Procedure	Applicable Factor
Storage (stock solutions)	$100\times$
Very simple wet operations	$10\times$
Normal chemical operations	$1\times$
Complex wet operations	$0.1\times$
Simple dry operations	$0.1\times$
Dry and dusty operations	$0.01\times$

Source: Reference 11, IAEA Safety Series No. 1, p. 23.

Volume 6000 of the Manual deals with construction, with Part 6200 being Engineering and 6300, Design Criteria. The meat of Part 6300 is in Appendix 6301, "General Design Criteria (Handbook)."[15] As the title would indicate, this is practically a blow-by-blow set of instructions for the engineers and well worth consultation by those concerned with building new radioisotope facilities or with modification of older buildings.

3.3 DESIGN PHILOSOPHY

A quick survey of apparent trends in laboratory and architectural design in general may be in order before specific consideration of radioisotope facilities. A laboratory at the beginning of the century was usually any available space in odd corners of the building large enough to set up a few tables or desks; the "services" being in the form of alcohol lamps, jugs of water and a few bottles of chemicals. The first major change in laboratory design came about when services—water, gas, electricity—became available in pipes and conduits. A separate room soon became designated as the laboratory, and these piped-in services were furnished at regular intervals along a fixed linear bench. While this change markedly increased efficiency and convenience, it also simultaneously restricted quick changes in internal arrangements and had the effect of creating a semipermanent installation, difficult and expensive to modify.[16] The bench approach naturally evolved into the modular concept, leading to standardization of furniture sizes, laboratory dimensions within a given structure, uniform patterns of services distribution and so on. This is the general situation today in most modern laboratories. Standardization has certainly increased flexibility in one way, since furniture and fixtures from one area can more or less be readily moved from one place to another. This flexibility is however still within the constraints established by the laboratory dimension, service and conduit piping, and ventilation-ducting framework.

The British[17] for some years have had an active program underway for developing laboratory designs and furnishings that would allow greater flexibility in making rapid rearrangements to meet changing conditions. Some of their ideas will be discussed in later sections.

3.3.1 The Overall Building   There were relatively few completely air-conditioned large buildings prior to World War II, and certainly not many of those were laboratories. New mechanical devices and engineering skills were developed after the war, and air-conditioning applications increased to the point where a nonresidential structure without a controlled inner climate is now a rarity in many areas. This development has had a striking effect on architectural design. The building became a sealed box in order to make the air conditioning work most efficiently, with many engineers arguing against the existence of any windows at all and probably with the secret wish that doors could also be eliminated if someone would only come up with some other approach to getting the occupants in and out. One effect of the sealed-box concept is that selection of the building site and the placement on that site have become thought to be less important design factors than formerly, a matter that could well receive reevaluation.

The completely controlled internal climate approach is a great user of energy. Stein, in his book *Architecture and Energy*,[18] argues that the pendulum has swung too far and that energy shortages and resulting skyrocketing costs indicate that a change is mandated towards a mixed design where the airconditioning would take on a secondary status and be used only on the unusual rather than on the normal day or only in certain portions of the building. He believes that careful consideration of site and orientation in most cases would allow whole sides of buildings to exist comfortably during certain seasons if windows to the open air were available; and that the use of overhangs, window "eyebrows" and similar architectural features would permit the same approach throughout many other parts of the structure—shielding the windows during the summer months but allowing the collection of solar heat in the winter.

The architects have a saying referred to as the two C's—"Conceive and Compromise." In other words, even the most satisfying of design ideas will almost certainly have to be modified when it encounters the real world of budgets, building codes, union rules, and materials limitations.

A third "C" might be added as a subheading under "compromise" in the case of radioisotope facility design, that of "contradiction" (or "conflict"). Requirements are not always mutually compatible, an example being the placement of staff offices. These should be as close to the work spaces as possible for convenience and efficiency, but this arrangement immediately runs head-on into the zoned-area concept described below.

3.3.2 Design Approaches Two fairly distinct strategies affecting design of radioisotope facilities have developed, termed by Garden[19] as "Concentrate and Confine" as compared to "Dilute, Disperse and Decontaminate." Garden and his associates at the University of California have been the chief proponents of the first concept and developed the "Berkeley box" (Figure 3.1), those little white-painted, plywood-constructed devices which have since spawned an astonishing number of descendants of all sizes and shapes (Figure 3.2), not only for handling radioactivity but also for manipulating viruses and for other similar applications. Radioactivity under the Concentrate and Confine philosophy is handled only in completely enclosed, restricted-volume, separately ventilated boxes; with all operations being carried out either through rubber gloves or with mechanical manipulators, both of which are essentially integral parts of the primary enclosure (Figure 3.3). The system has the advantages of decreasing the required air flow in a drastic manner, of ensuring that the location of the radioactive material is known at all times and of allowing somewhat more "normal" design and activities in the laboratory itself. The chief disadvantage for the radiochemist is that working volumes are very restricted, making it necessary to plan and check every detail of an experiment in advance and

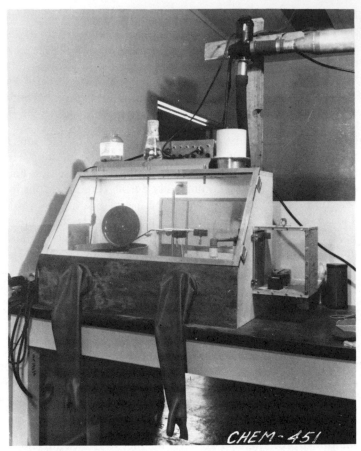

Figure 3.1   The original "Berkeley" box. (Courtesy E. K. Hyde, Lawrence Berkeley Laboratory.)

making it difficult to introduce substantial deviations after the operation has once begun. The restricted space also noticeably slows down the necessary manipulations, increases the chance of accidentally upsetting equipment, and, on occasion, leads to problems of dispersing acid fumes or high heat loads. This system has attained practically universal acceptance for handling alpha emitters, but very little application in dealing with the purely beta-gamma nuclides. (On the hot-cell level the two radioactivity-handling strategies tend to merge since many existing cells are in effect heavy-walled boxes, although not necessarily air tight. As an aside, if the time required to carry out a particular operation with nonactive materials on the open bench is assigned a value of 1, the corresponding number for radiochemistry hood work is probably around 1.2; for gloveboxes, in the 1.5 to 3 range; and for hot-cell operations, seldom less than 3 with the upper limit very indefinite, depending on just what has to be done.)

Figure 3.2 Berkeley box descendants. (ANL Photo 121-3062.)

The Dilute, Disperse and Decontaminate title of the second handling technique is a bit misleading in suggesting a rather casual approach, certainly not present in any properly designed and managed radioactivity handling facility. The approach does allow much more open operations and is characteristic of most non-alpha laboratories today.

The whole laboratory to a much broader extent is considered as being the working area. The rate of radioactivity release to the environs is made zero or controlled to nonhazardous levels by air exhaust filters, liquid waste hold-up tanks, limited-access control devices, etc.; all part of the building itself rather than being directly associated with the primary enclosure. The system does tacitly assume that occasional internal contamination will occur and necessitate removal as quickly as it is discovered. While the primary defense weapon of the Concentrate and Confine philosophy is the glovebox; those of Dilute, Disperse and Decontaminate are the radioactivity hood, overall laboratory design, intensive monitoring and administrative control. Most of these latter features are in actuality applied to both approaches so the two tend to merge in practice in spite of some real differences in the philosophical attack.

3.3.3 Zoning On a more concrete level, even for the Concentrate and Confine approach, the most important principle in radioactive facility design is that

Figure 3.3   Shielded box with tong-type manipulators. (Courtesy Kewaunee Scientific Equipment Corp. Adrian, Mich.)

of zoning of areas, although the extent to which it is necessary to carry out the gradation pattern to full completion will depend on the levels and types of active material handled and on the size of the operation. Active work should be concentrated in controlled areas, with access through increasingly suspect regions, that is, from cold to lukewarm to warm to hot. The IAEA[11,12,20] has formulated this idea in more specific terms, a classification widely quoted and accepted. Their definitions are not always consistent (in reverse order in SS No. 38), but basically are as follows:

Zone 1   No radiation or contamination hazard exists in this area. Entry is unrestricted if radioactivity is the only criterion, although access may be limited for other administrative reasons. (Offices, conference and lunch rooms, libraries, ordinary shops, certain corridors and sanitary facilities, etc.).

Zone 2   Areas in which the average external radiation level is not greater than 0.1 R/week. Contamination is highly unlikely and assumed to be absent, although the space is frequently monitored on a routine basis. Access is limited to radiation workers but without any special clothing regulations. (Clean side of change rooms, buffer corridors, isolation rooms, external hot cell operating areas, counting and control rooms, etc.)

Zone 3 Areas in which potentially damaging external radiation levels could exist and contamination is always a possibility. Access limited to radiation workers wearing dosimeters and special clothing. (Laboratories, hot-cell transfer areas, working side of change rooms, filter and active sample storage rooms, hot shops, etc.)

Zone 4 Areas in which the external radiation and contamination levels are or could be high. Access is rigidly controlled with special clothing, close monitoring and other precautions mandatory before entry. (Hot cell, radioactive hood and glovebox interiors; highly contaminated areas of any type, etc.)

The physical arrangement of the facility should be such that increasing control and surveillance of personnel and materials can be imposed in going from Zone 1 to Zone 4. As indicated, access to Zone 1 can be completely free from the hazard point of view. The 1-2 barrier should be somewhat restricted, the 2-3 very definitely so, with change rooms, scanning monitoring devices and so on for large operations. Zone 4 spaces should be entered only under the most stringent control conditions. There may have to be several 2-3 interfaces in a large facility having both laboratories and hot cells or engaged in a number of different operations.

Unruh[21a] suggests that for large operations the health physicists should have their quarters between Zones 1 and 2, including space for storage of monitoring equipment and supplies and their own laboratory for counting air filters and equipment smears, in addition to office facilities. He recommends that external radiation levels in Zone 1 not exceed 0.2 mR/hr, and not over 0.5 mR/hr in Zone 3, to which no direct access from outside the facility should be permitted. Radioactive material transfers should never pass through Zone 1 in either direction from Zone 3; movement of any items at all from Zone 3 to 1 (notebooks, etc.) should be held to a minimum with careful monitoring at the 2-3 interface.

ISO/R1710[14] also adopts the zone concept and specifies the background radiation level in each area as fractions of the ICRP annual limit: Zone 1, exposure such that the yearly dose is under 0.3 times the limit; Zone 2, 0.3-1.0 times the limit; Zone 3, up to the limit; and Zone 4, not routinely entered. They also suggest that the zones be referred in terms of colors: 1—may be colorless, 2—green, 3—yellow-orange (reduced to amber in other references) and 4—red. The designation of white would be reserved for dustfree "clean" rooms. (Others would use white for areas outside the facility beyond the range of any possible contamination.) Materials transfer into the higher-numbered zones should be by routes other than those used for routine personnel entrance and exit.

3.4 FACILITY LAYOUT

A new laboratory in an old building need not be too much of a problem however if only low levels of short-lived beta-gamma activity are to be utilized. The IAEA Type C-laboratory definition would apply, and quite possibly the only additional precautions that would be needed would be for adequate monitoring equipment and a system for holding liquid and solid contaminated waste until the radioactivity had died away. If longer-lived activities are to handled at higher levels, the laboratory should preferably be as isolated as possible from the rest of the building and have an independent entrance so that shipping containers and waste would not have to be moved through the nonactive portions of the structure. Installation of separate ventilation and of exit filters for the exhaust air might be required before an operating license could be obtained. In a multistoryed building, a location on the upper floors would reduce the length of potentially contaminated ductwork between the hoods and the exhaust stack, but against this advantage would be the question of moving active materials through the lower floors unless the space chosen had its own separate elevator. Floor loading could also become a problem if there was a need for even temporary shielding or for moving heavy shipping casks in and out. Pipe runs would also be much longer if new services had to be installed. On balance, a location on the ground floor as near as possible to a (preferably dedicated) loading dock would seem to be the best choice if the ventilation problem was not thereupon made insoluble.

The designer of special space in a new building would obviously have a much easier job, particularly in establishing buffer zones. The laboratories would best be in a separate wing, no matter how small; have entrances of their own, one giving easy access to a private, even if rudimentary, materials receiving and shipment dock; have their own ventilation and laboratory liquid-waste disposal systems; be buffered from the inactive areas; and be placed so that the exhaust from the ventilation system could not be drawn into the wing's own air supply, that of the rest of the building or those of any other nearby structures.

3.4.1 Auxiliary Rooms
A radioisotope facility cannot operate without access to activity measuring devices, so the problem is that of their location. Contradiction again comes into play—the counters should be as close to the operation as possible for the sake of efficiency; but this may place them in areas where corrosion could occur or where intolerable backgrounds from the point of view of accurate counting exist or where maintenance of the instruments would be difficult. One answer in a suite of laboratories would be to apply the zoning concept within the complex itself—high-activity laboratories at one end and the counting room at the other—with a gradation in permitted activity levels in between. A few counters could be provided in the hotter areas for quick evaluation of ongoing experiments with the thought that more precise data could be subsequently obtained in lower background counting facilities if

necessary. If such a zone-within-a-zone plan was adopted, hand-and-foot counters should at a minimum be placed at the entrance to the high-activity laboratories as well as at the Zone 2-3 interface, and all intervening areas frequently monitored.

One or more modules adjacent to the counting room could also be designated as cold, even if embedded in the Zone 3 area. These would be used for reagent preparation, glassblowing, computer terminals, communal writing desks for the staff, etc. (The activities listed might not all be compatible in a single room.) Several rooms in a large group of laboratories should also be reserved from the start for instruments other than counters (chiefly spectrophotometers of various types) and deliberately designed with a maximum of open area, special services if needed and a modest amount of drawer and cabinet storage space. A single hood with adjacent single bench module might be advisable for loading and unloading of samples, with a limited amount of bench space elsewhere for instrument repair and maintenance activities.

Modern analytical and microbalances are much more rugged than formerly, so the need for a separate balance room is questionable. Dish washing in a large operation might be centralized in a separate module, a room which should also have facilities for storage and redistribution of the cleaned items. Flammable solvents are of course a major storage problem in any laboratory, but particularly so when radioactivity is being handled because of the very real danger of contamination spread in case of fire. In the rather idealized type of facility being considered here, a separate highly ventilated cubicle for solvent storage, having explosion-proof lighting fixtures, automatic heat detectors, and fire-fighting equipment would be highly desirable. The alternative is an adequate number of OSHA-approved solvent-storage cabinets placed with careful avoidance of the areas having high concentrations of potentially dispersible active materials.

Radioisotope work also requires much larger quantities than ordinary of dispensable items such as inexpensive glass- and plasticware, paper towels and tissues, cartons, labels, rubber gloves, disposable shoe covers, etc. An undesirable amount of personnel traffic between the controlled radioactivity areas and other parts of the building will result if such items have to be obtained from a central stockroom. This is a big argument for a depot of these fast-turnover materials in a cold module in the controlled area, even if without a custodian. Two or three adjacent small cold modules in the Zone-3 area might well be reserved in a new facility for handling the chemicals, supplies and solvent-storage problem.

Most new or existing laboratories will not have the space or the budget to separately house all of the varied activities outlined above. The designer should nevertheless remain aware that counting, instrumental measurements, data recording, weighing, glassblowing, dishwashing; and storage of glassware, chemicals, supplies and solvents are activities common to most radioisotope operations no matter what the size and that the less compatible functions should be separated as much as possible in whatever space is available.

3.4.2 Active-Sample Storage A problem in the radioisotope field is that of housing active stock solutions, sealed sources, and samples and fractions gener- ated during experiments where at least short-term retention is desired for later possible confirmation purposes. This problem in recent years has been compli- cated by the need for security provisions ("safeguards") in addition to the strictly safety aspect, particularly if the work involves fissile materials or com- pact sources of intense radioactivity of possible appeal to a sick mind. Both the safeguards requirements and the need to maintain an up-to-date registry of all active samples for safety and accountability purposes strongly favor the estab- lishment of a centralized storage unit for those active materials not used every day. This facility should be placed under the supervision of a competent custo- dian and alternate whose decisions on withdrawals and additions are final.

There are of course arguments in the other direction. A central storage unit implies more traffic of active materials through the laboratories, with the atten- dant chances for mishaps along the way. The arrangement is also less efficient from the researcher's point of view, leading to a quite reasonable resentment against having to spend an appreciable fraction of each day in depositing or extracting frequently used materials from storage. A possible compromise is to keep the bulk of the less-used materials in a central unit (a locked room), but then to furnish moderately shielded, ventilated and lockable safes or cubicles in the laboratory for items used more frequently.

Elaborate storage facilities are of course not required in many radioisotope laboratories, particularly if the work does not involve materials requiring safe- guards protection. The shipping container in which an active stock material is received is in many cases nonreturnable and thus can be placed in a secure corner in a hood to serve as a shield until the stock is all used. Simple lead, iron or concrete "pots" are commercially available or can be rather easily fabricated to serve the same purpose. Ward[6a] suggests a large concrete block having holes of different sizes and depths cast or drilled into the top as a storage unit. Plugs adequate to furnish the needed shielding and having eye-bolts or other devices to simplify tong or mechanical handling are used to cap the holes. An old commercial safe might be adequate for alpha emitters. A shielded wall unit having a large number of cubicles can be custom built if a variety of different isotopes is to be routinely handled and cross contamination is a concern. Such units should be individually vented to the exhaust runouts to avoid drawing activity into the room due to the aspirating effect of opening a cubicle door.

Section 10.2 of volume III of the *Engineering Compendium on Radiation Shielding*[22] contains considerable information on shielded storage units. IAEA Series No. 1[11] discusses sample storage, although primarily from the operation- al rather than the facility point of view. The older compilation by Stang[23] of hot laboratory equipment describes various storage units.

The matter of safeguards is of course predominantly a problem for nuclear fuel cycle facilities and is receiving much current attention including various international symposia.[24] Unruh[21a] briefly considers the safeguards problem for facilities handling gram amounts of plutonium, and ERDAM Appendix 6301 (Part II, Sec. 1.16)[15] gives engineering guidances for safeguards installation.

3.4.3 Area Specifications The same ERDA manual (Part II, Sec. B) essentially mandates a modular approach to laboratory building design for its engineers, and the modular concept is undoubtedly basic to most structures of the type being built today. A standard dimension is used repetitively throughout the building to determine the position of walls and partitions, duct run-outs, air supply and heating units, and service piping and conduits. The module length chosen will also affect the position of the building's vertical support members, crossbeams and joists; the placement of doors and windows; and the lengths chosen for the furniture and hood units. The standardization of so many items throughout the building obviously offers opportunities for economy during construction and facilitates changes after occupancy. The concept may not be all pure gold as it can lead to unnecessary overdesign in some areas unless care is taken.

Ruys[10a] defines modules as "basic" and "planning." A good example of the former is the use of 4 in. as the normal base in the home construction industry where most dimensions are divisible by four and studs and joists accordingly put on 16-in. centers. Ruy's "planning" module is a three-dimensional box used repetitively as far as possible all through the facility.

Mellon[9] has published a recent group of articles in which he surveys existing and planned university and college chemistry and science buildings. In the first of these papers he speaks of the various modular widths seen in the course of his review. These varied considerably, but were usually 4, 4 ½, 5, 5 ½ or 6 ft. (He specifically mentions the Mies van der Rohe-designed science building at Dusquesne University where the basic module is 7 ft in either direction, thus bringing the vertical support columns out on 28-ft centers.) Solomon[6] in an earlier similar review of industrial laboratories found 8-, 9-, 10- and 11-ft modular widths, with the preponderance of laboratories being 18–22 ft in width. Depths varied all the way from 19 to 52 ft, but most were in the 22–28-ft range.

ERDAM Appendix 6301 strongly endorses 12 ft × 24 ft or 24 × 24 ft for individual laboratories, although with a few qualifying statements because of the impossiblity of making the same shoe fit every foot. The 12-ft module as a base has advantages. The most readily accessible off-the-shelf commercially available laboratory furniture items are usually 4 ft in length, a number that fits nicely into the base of 12. A 12-ft module (as compared to 10-ft module) also provides wider aisle space between the laboratory benches, a desirable feature in radioisotope work where heavy transfer casks and monitoring equipment of various types must be moved in and out on occasion.

Solomon in his survey not only found considerable variation in laboratory sizes, but also in the amount of space calculated per occupant, the numbers running from 114 to 220 ft^2 person, the most common area however being in the 140–180-ft^2 range. At the end of his paper Solomon presents a number of possible single laboratory layouts and indicates the number of persons each is designed to house. Since his interest was primarily in industrial laboratories, many of his plans show partioned-off offices in the laboratory itself, a dubious practice in even a medium-level radioisotope handling facility.

It is of course not too surprising to learn that the population density of laboratories shows considerable variation since different organizations have different programs and aims. Office space should however presumably be somewhat more predictable. ERDAM Appendix 6301, Part II, Sec. 4a recommends:

Office Type	Population density (ft²/person)
Executives (Division Director and above)	200
Scientists, engineers	100-150
Secretaries	60- 75
Clerical, steno pools	60
Clerks and files	75

For conference rooms where up to thirty people might assemble around tables, the ERDAM allowance is 20 ft²/person, reducing to 8 ft²/person for auditoriums, seminar rooms, etc. Draftsmen are allotted 75 ft²/person, exclusive of space for storage of drawings. The executive allowance can go up to 400 ft² if the individual holds many meetings in his office. (These numbers of course are for planning purposes only.)

Corridor widths again depend to a certain extent on use and circumstances. ERDAM Appendix 6301 specifies a clear width of 6 ft for the main corridors of administration and office buildings if the doors into the adjoining rooms open inward. The manual allows 5-ft-clear width in light-traffic areas if the corridor and door openings meet the standards of NFPA (National Fire Protection Association) 101, "Life Safety Code."[25a] A width of 7.5 ft is approved for laboratory corridors if the doors swing outward into the traffic pattern.

Everything else remaining equal, widening of a corridor correspondingly expands some horizontal dimension of the building shell with attendant increases in construction costs. Thirty plus years ago it was estimated[26] that after the expensive external framework and the corresponding internal support structures had been covered in the budget estimates, add-on space such as a wider corridor would be at about ⅓ the overall cost rate.

3.4.4 Structural Specifications Specification of floor loading should be as realistic as possible since the number chosen will have a material effect on the size and the cost of the building structural members. Office building floor loadings are usually in the 100-150-1b/ft² range. A figure of 200 1b/ft² was used for the Argonne Chemistry Building, but at the time of design of its large hot laboratory addition (Wing M[27]) it was felt that this was marginal for a facility of its type so 300 1b/ft² was specified. The British[28] established a figure of 2000 1b/ft² for the chemistry building at Harwell in England. Achieving such a high value was not however too difficult in their case since the laboratory areas are on grade.

Concern about the safety of nuclear-fuel-cycle facilities has in the last few years generated numerous reports and analyses dealing with the behavior of structures during natural disasters such as earthquakes and tornados. Nuclear reactors and fuel reprocessing plants of course receive the brunt of this concern, but laboratories handling high levels of activity, particularly plutonium and other transuranics, are also receiving attention. This could affect planning of a new building since design and construction of a completely earthquake- or tornado-proof building is an expensive business. The Nuclear Regulatory Commission has issued[29] several guidelines, primarily for the industrial side of the nuclear fuel cycle.

3.5   THE HVAC SYSTEMS

While the heating, ventilation and air conditioning (HVAC) associated topics are discussed here under laboratories, most of the material applies equally as well to the shielded facilities discussed in the next chapter.

The most influential organization in this country in establishing standards and procedures in ventilation matters is the American Society for Heating, Refrigerating and Air Conditioning, (ASHRAE).[30] In addition to publishing standards, data and guide books the Society also brings out a yearly "Handbook and Product Directory" with the subject emphasis in each being rotated on a cyclic basis. Thus the ASHRAE handbooks for 1974 and 1978 were on Applications; in 1975 on Equipment; 1976, Systems; and in 1977, Fundamentals. The material of most interest to the present section is Chapter 15, "Laboratories", of the Application handbooks of 1974 and 1978. ASHRAE Standard 90-75, "Energy Conservation in New Building Design" is also pertinent as are previously cited portions of ERDAM Appendix 6301. The 1976 revision of Part I, Section D, of this last document lists a number of other applicable standards, codes and guides.

3.5.1   Air Supply and Exhaust   The purposes of a structure's HVAC system are to bring in an adequate quantity of ventilation air, to clean it if necessary, to adjust its temperature to satisfactory levels, to circulate it through the specified space and eventually to exhaust it back to the atmosphere after cleaning it of any noxious materials picked up along the way. (Some or all of the air may in suitable cases be recirculated before this pattern is complete.) The HVAC system is primarily for supplying breathable air, for internal climate adjustment and for removal of excess heat and unpleasant fumes. The ventilation in a laboratory handling dangerous substances such as radioactivity is however in addition possibly the most important of the basic safety protection tools (along with shielding) and will accordingly receive extended consideration here.

Different HVAC philosophies, not all equally suited for nuclear facilities, are used in various applications. Each single laboratory or other space can be furnished with hardware to allow it to be heated and ventilated as a separate unit. This improves flexibility considerably since individual regions can be shut off when not in use, but adds to the capital and maintenance costs. The opposite approach is to service a large group of areas with a centralized air-handling and treatment facility (unitary versus central supply). The air supply-exhaust system can be such as to allow variations with time in the quantity of air furnished, again permitting energy savings by drastic cutting of the supply during periods when the space is unoccupied. The opposite approach utilizes a system where the air supply and exhaust are constant—more costly but much easier to keep in balance (variable versus constant supply). All of the air can be brought in, put through the system and then exhausted; or some of it may be recirculated to reduce energy losses (recirculated versus once-through). Each laboratory or module may have its own individual exhaust stack—expensive, but giving flexibility since only a small portion of the building is shut down

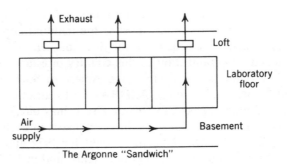

The Argonne "Sandwich"

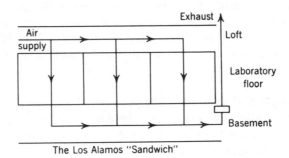

The Los Alamos "Sandwich"

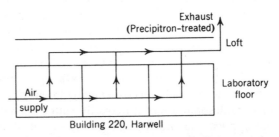

Building 220, Harwell

Figure 3.4   Three laboratory air-handling systems.

during repairs—or all of the reject air may be collected and expelled through a single large stack (single room exhaust versus common stack).

The conditions in a radioactivity-handling facility effectively limit some of the choices. Because active materials in the hoods, cells and storage facilities must be guaranteed to remain there with no chance for even momentary reversals in air flow bringing them back into the surroundings, and because recirculation of even elaborately cleaned air is a bit chancey in case of equipment failure; the general pattern in radioactivity-handling facilities has been one of centralized and constant air supply on a once-through basis.

Examples of some air-handling systems are diagrammed in Figure 3.4. In the Argonne "sandwich" the air for each wing is brought in at the service floor level, conditioned and filtered, distributed through the zoning pattern to the separate laboratories on the main floor (as shown in Figure 3.5) and exhausted through the filters and stacks which occur at each module position in the fan loft. A downdraft system is used in the Los Alamos sandwich—the air is supplied and conditioned in the attic, then distributed to the laboratories from which it is drawn downwards through the hoods and boxes, collected in a common duct in the basement and carried to a common stack for discharge. The Harwell system is updraft, but in this case the exhausts from groups of laboratories are routed to a common duct for transfer to a single stack. All three systems are of the once-through types. (Figure 3.4 applies to the original buildings at all three sites and may not be typical of any newer structures.)

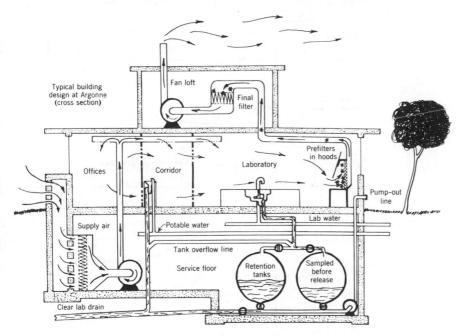

Figure 3.5 Typical laboratory air and water supply and exhaust patterns. (ANL Photo 120-5964.)

The main purpose of a high stack is to obtain maximum dilution of contaminants in the exhaust as quickly as possible. A tall stack also simplifies the problem, mentioned previously, of being certain that the exhaust is not drawn back into the building's own air supply or that of any neighbors. Figure 3.6 illustrates the patterns produced by stacks of different heights. In all cases the logical place for the building air inlet would be on the side towards the prevailing wind.

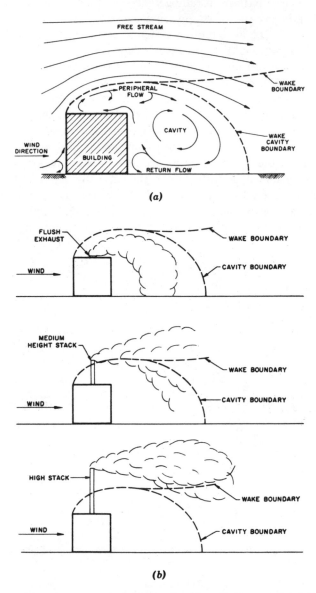

Figure 3.6 (*a*) Air-flow pattern around a building; (*b*) effect of exhaust stack height. (From Refs. 30 and 32, Courtesy ASHRAE, reprinted with permission.)

Figure 3.6 shows the behavior of smoke plumes discharged from single buildings. In the case of a short exhaust stack, the wind pushes the plume down towards the ground, leaving a space adjacent to the structure (the cavity) in which there is much internal turbulence and mixing. If the stack is somewhat higher, the cavity volume decreases and part of the plume is carried off directly by the wind. If the stack is high enough, all of the exhaust goes into the free wind stream where it is highly diluted by diffusion and mixing. Where several buildings are adjacent to each other the problem of locating the individual air intakes becomes more complicated as shown in Figure 3.7. Wind-tunnel tests with simple models may be indicated in such a situation.

Air intakes should not be located near loading docks or parking areas, particularly in colder climates. Truck drivers like to leave the motors running in their vehicles while making winter deliveries and the exhaust fumes can quickly permeate an entire building if the air intake is in the wrong place.

Design of the ventilation pattern for radioactivity control has a very direct and fundamental relationship to the zoning concept discussed in Section 3.3. The air must always move in the same direction—from clean to more suspect areas and then to possibly dirty regions. From there, after decontamination treatment (usually by filtering), the cleaned air is exhausted to the atmosphere. The pattern is thus from Zone 1 successively through Zone 4, clean-up and out as shown in the simplified sketch of Figure 3.5. Outside air is drawn into the building, filtered to remove dust, cooled to squeeze out excess moisture, adjusted in temperature and then transferred to Zone 1 areas. From there it is drawn through the buffer areas into the laboratories, exhausted through the hoods and gloveboxes, decontaminated by filters and discharged to the atmosphere. A small decrease in pressure is maintained in each successive zone as compared to the one previous in order to maintain the air flow direction. The whole building is often also kept at a slightly negative pressure as compared to the outside so that any leaks will be inward and any possible contamination release thus confined. Maintenance of these pressure differentials requires a careful balancing of the air volume introduced into and exhausted from each zone. Since the pressure-drop pattern in the system will be affected by many factors over a period of time (dirty filters, slipping fan belts, etc.) the whole system has to be generously instrumented so that continuous adjustments are

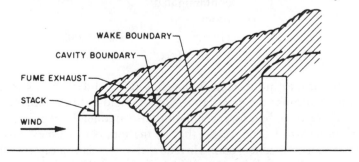

Figure 3.7   Air-exhaust patterns around adjacent buildings. (From Refs. 30 and 33, Courtesy ASHRAE.)

made automatically. Air-locks are often used between zones as a means of obtaining better control of the pressure situation in each area. These locks are small entrance lobbies having doors at each end electrically interlocked so that only one can be open at a time. In principle, airlocks can be used to isolate one portion of a building, thus allowing that space to have its own independent air supply and exhaust system.

Pressure differentials are expressed in a confusing number of different ways in the ventilation literature (everything but the official SI unit, the pascal), but probably most frequently as inches of water. Some equivalents (rounded to three figures) are:

1 in. of water equals:

25.4	mm of water
0.0738	in. of mercury
1.87	mm of mercury
1.87	torr
2.49	millibars
0.00246	atmospheres
249	pascals (newtons/meter^2)

Using these equivalents to convert a sampling of the various recommendations appearing in the literature to inches of water gives the data shown in Table 3.3.

(S. J. Vachta,[34] a ventilation engineer with many years of experience, believes that some of the numbers shown are unrealistically high. He would recommend:

High side	Low side	Differential (inches water)
Outside	Zone 1	0.
Zone 1	Zone 2	0.05
Zone 2	Zone 3	0.10
Zone 3	Hot cell or contaminated room	0.5–1.0
Zone 3	Hood	0.003–0.005
Zone 3	Pu glovebox	0.5

Vachta points out that in the Zone-3-glovebox case a differential of more than $\frac{1}{2}$ in. of water will cause the gloves to become so rigid as to be unmanageable.)

The literature on laboratory ventilation also frequently expresses recommendations in terms of changes/hr, that is, the number of times during that period that the air in a zone, room, glovebox, cell, etc., is completely replaced. In a constant-supply-constant-exhaust system some of the incoming air can be

Table 3.3 Zone Pressure Differentials

High Side	Low Side	ΔP (inches of water)	Reference
Zone 1	Zone 2	≈ 0.1	14
Zone 2	Zone 3	0.2−0.4	
Zone 3	Zone 4	0.6−0.8	
Zone 2	Zone 3	0.1−0.25	21a
Zone 1	Zone 4	1.25	
Outside	Zone 2	0.08−0.2	21c
Outside	Zone 3	0.4−1.2	
Outside	Zone 4	1−2	
Outside	Zone 1	0	21d
Outside	Zone 2	0.1−0.2	
Outside	Zone 3	0.4−0.6	
Outside	Zone 4	1.0−1.6	
Outside	Zone 1	0	21b
Outside	Zone 2	0.1	
Outside	Zone 3	0.2	
Outside	Pu boxes	0.6−1	

immediately routed as a supplemental supply to the higher-numbered zones (where more changes/hr are needed) without all of the intake having to pass through Zones 1 and 2.

Table 3.4 presents some air change recommendations from the literature. Some of the variation seen in the numbers is probably due to the fact that different authors make different assumptions as to the components of a zone. The Czechs[21b] for instance call their gloveboxes Zone V and the laboratory in which the boxes are located is classified as Zone IV.

ERDAM Appendix 6301 does not make air change recommendations on a zone basis, but does however state that 5 scfm (standard cubic feet per minute) of air should be furnished per person for normal respiratory and odor control in offices, auditoriums and similar spaces where light smoking may occur, that is, Zone-1-type areas. Hughes[37,38] considers the air change question, but uses only quoted figures in presenting specific numbers.

A decision on the number of air changes needed in each zone obviously is one way of determining the total quantity of air that will be needed to ventilate a facility. There are other factors however that must be considered. Thermal loads are normally not a problem in radioisotope installations because of the high volumes of air needed for safety purposes, but the engineers will still have to evaluate the heat removal situation to be certain that there are no unusual circumstances. Three factors are considered—load, use and diversity.[30] The load factor is derived by listing the equipment in a particular area and calculat-

Table 3.4   Air Change Recommendations

Volume	Recommended Changes/Hr	Reference
Zone 1	2−4	21d
Zone 2	2−5	14
	12	21c
	6	21b
Zone 3	20	21b
	8−15	21a
	10	14
	>12	35
	6−15	21b
	5−8	21d
Zone 4	>6	21d
	"Some 10's"	14
	20−30	21b
Hot cells	60	14
Gloveboxes	15−20	21b
Ordinary lab	4−8	36
Toxic Products lab	Up to 30	36

ing the aggregate heat release from manufacturers nameplate information. (Equipment in hoods or separately cooled is not included.) The use factor relates to the fraction of time in which a particular equipment item in expected to be in use. The diversity factor is an evaluation of the number of pieces of equipment in the room likely to be in simultaneous operation. Application of these factors then allows computation of the heat load that must be removed by the air flow in addition to that generated by the room occupants and the lighting. ASHRAE Chapter 15[30] gives some usual ranges for a 250-ft^2 laboratory module:

Equipment Loading	Heat generation
Normal	15–30 btu h/ft^2
High	30–60
Very high (electronics)	70–140

The figures represent net heat gains to the space after deduction for diversity and hooding.

A probably more critical factor in a radioisotope facility for establishing the total air supply need is however the decision on the number and types of hoods that will be needed and the amount of air that will have to be furnished to each of these exhaust units. (The number of gloveboxes is much less controlling because of their very substantially lower air requirements.) In effect, the critical parameters are the total maximum area of hood face opening expected during operation and the design velocity established for the air passing through that opening. (This velocity specification is usually given as lfm, linear feet per minute; fpm, feet per minute; or m/sec, meters per second. 1 m/sec is approximately equal to 200 lfm. Air volumes are most often expressed as cfm, cubic feet per minute.)

3.5.2 Hoods Radioactivity hoods vary as described below, but if one is chosen furnishing 12 ft^2 of maximum opening and if an average face velocity of 125 lfm is established for design purposes, the hood will require 1500 cfm of air supply, If it can be guaranteed that the hood opening will never exceed 8 ft^2 when radioactivity is present, the air demand is reduced to 1000 cfm. In a laboratory containing a number of hoods an estimate of the average face opening per hood expected at any one time can of course be taken and multiplied by the number of hoods to calculate the total supply of air needed for the room. An alarm system indicating that the overall allowable face opening is being exceeded is almost a necessity in such a situation.

The 125-lfm face velocity figure quoted above ($\approx$ 0.64 m/sec) is a not unreasonable average for most radioactivity hoods. Hughes[37] suggests 100 lfm for ordinary chemistry laboratory fume hoods, but states that the British Ministry of Labor has recommended 400 lfm when working with tritium gas in order to prevent back diffusion. A design value of 150 lfm was chosen for the Wing M installation.[27] (Maintaining an open-burner flame in the hood is difficult and lighter objects tend to go with the wind if the velocity gets much above this figure.) The specification has as its main purposes those of being certain that the air flow is always from the operator towards the radioactivity in the hood and of preventing that material being drawn back into the room by stray air currents or the aspirating effect of opening a nearby door, the passage of personnel walking by the hood or the movements of the operator's arms as he works in the unit. Air velocities will not be exactly uniform at all levels of a wide-open hood face but the minimum at any spot should never be less than 80% of the design value. Because maintenance of an adequate air velocity at all times is such a critical safety feature, standby exhaust fans operated by an emergency power system should be available to immediately take over in the event of failure of the main power supply.

Radiochemistry hoods are of course a direct outgrowth of the ordinary chemistry laboratory fume hood (or fume cupboard, as out British friends would have it), but with additional attention to design details in order to minimize eddying and with added controls to be certain that the air velocity will not be abruptly changed when the hood sash is opened or closed. Figure 3.8 pictures one of the currently available commercial versions. The hood shown is typical of most of the units now being manufactured and is based on designs worked out at the Oak Ridge Institute of Nuclear Studies[4] and at Lawrence Livermore Laboratory[39] in the early 1950s. The chief visible distinguishing feature is the "air-foil" treatment of the face opening; a shaping of smooth contoured surfaces around the face to reduce eddying. The hood structural components are also strengthened to carry the weight of isotope shipping containers or temporary lead brick shielding, the exhaust exit area is contoured and the baffling slots are modified to handle high air flows. Many of the commercial units have 304 or 316 stainless-steel interiors with a minimum of welded seams and perhaps with coved inner corners to simplify decontamination. (If much hydrochloric acid is to be used, painting the inner surfaces with a corrosion resistant paint may be advisable. Units so pretreated are commercially available.) Utility outlets are furnished towards the front within the hood, but the controls are located on the outside in order to reduce the need for the operator to insert the upper part of his body into the hood itself.

Figure 3.9 shows a laboratory utilizing a different style of radiochemistry hood modified[40] from a design developed by the Blickman Manufacturing

Figure 3.8   Radiochemistry hood. (Courtesy Hamilton Industries, Two Rivers, Wisc.)

Company and widely used in several of the national laboratories. The "Blick-man" sacrifices the air-foil approach but offers several advantages in return. The upper panel of the face is a single pane of glass normally fixed in place, although it can be removed (or raised on tracks in the original version) if the entire face has to be opened for equipment building or installation. The lower face may be left completely open for cold or low-level work or may be fitted with a panel containing either sliding glass doors or glovebox rings. In the sliding-door case, the use of three glass panels each in its own separate track means that no more than two-thirds of the face can be open at any time. Three panes in two tracks cuts this maximum opening to one-third, whereas four panels in two tracks would allow a 50% maximum. This feature, which is sometimes a nuisance to work with, offers the ventilation designer a positive method for controlling the maximum possible overall area of open hood face in the room.

The modified Blickman unit seen in Figure 3.9 can be used as a variable-opening hood, as a glovebox (by fitting a panel over the bottom opening) or the lower portion can be left entirely open for lower-level work. Fiberglass-reinforced polyester resin was used as the construction material for the hood and also for the room ductwork. This material has worked out very well, being corrosion resistant and easily decontaminated if any spilled activity is removed

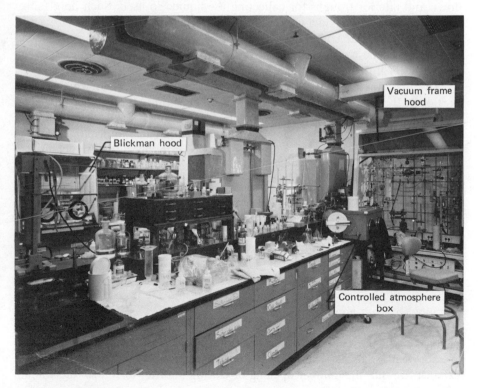

Figure 3.9 Radiochemical laboratory showing ventilated enclosures. (ANL Photo 121-6533.)

reasonably promptly. The contamination otherwise tends to dig into the surface, probably by the recoil mechanism mentioned elsewhere.

The smalled gloved unit at the right of Figure 3.9 can be ventilated either by a low flow of specially dried air or by an inert gas, thus serving as either a dry- or a controlled-atmosphere box. The large unit at the very back and to the right of the figure is a modified "California type" or "vacuum-line" hood.[41]

The hoods that have been discussed are of the up-draft type, that is, the air supply to the hood is exhausted through the top of the unit and carried out of the laboratory by overhead ducts. Down-draft hoods, where the pattern is reversed, have not been applied to the same extent for radioactive work. The chief advantage argued for the down-draft design is that it permits washing down of the hood interiors and the ductwork for decontamination purposes, and that the exhaust air itself can be washed with sprays for removal of active particulates. The design however creates other problems. The wash or spray water must be collected and disposed of as active waste and the exhaust air may have to be dried to a certain extent before reaching the HEPA filters in order to prevent their clogging and deterioration. The washdown principle has one application where there are clear advantages and that is in situations where large quantities of perchloric acid must be handled. Such "perchloric acid" hoods are commercially available and, in addition to the down-draft feature, are characterized by all-metal construction, built in sprays to facilitate washdown and interior troughs for collecting the liquids so used. Such hoods must be vented by all-metal ductwork, and the problem of preventing any of the acid reaching the organic components of the final filters used in radioisotope facilities must also be considered. Most radiochemical procedures fortunately call for relatively little use of perchloric acid.

The manufacturers produce "auxiliary air fume hoods" in which only a portion of the needed air is taken from the room supply, the balance (5%–70%) being brought in directly from the outside or from other parts of the building. Units of this type have found very little application in radioisotope work in spite of the potential for energy savings. This is probably due to the complications that would be introduced into maintaining balance in an already complex air handling system.

There must be some mechanism in a constant-supply/constant-exhaust system for handling the excess air when the hoods are not open to their full capacity. This is generally handled in one of two ways. In the first the air not needed by the hoods is exhausted through a room by-pass, that is, it is routed directly from the room into the exhaust ductwork. (The rectangular grill in the ceiling of the laboratory pictured in Figure 3.9 is such a by-pass.) In the second technique the by-pass is part of the hood itself and the units are termed "proportional by-pass hoods."

Figure 3.10 diagrams three different hood types. Figure 3.10(a) is an example of the type of hood that would be used in the room by-pass arrangement. A pressure sensing device within the hood actuates a motor-driven damper in the exhaust run-out, the damper opening or closing in order to control the amount

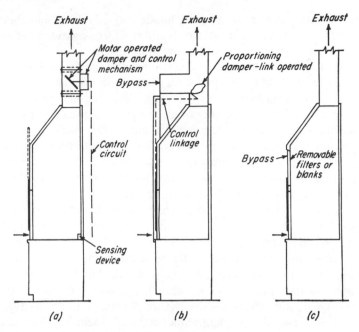

Figure 3.10 Hood designs for air-velocity control.

of air entering the hood under varying face-opening conditions, thus maintaining the desired air-intake velocity. Figures 3.10(*b*) and 3.10(*c*) are both proportional by-pass types. In Figure 3.10(*c*) raising the sash covers part of the by-pass opening, thus proportioning the air intake between the hood interior and the direct release to the exhaust. When the sash is completely open the by-pass is essentially covered; when the sash is closed, by far the bulk of the air exits through the by-pass. The sash does not cover the by-pass opening in the design of Figure 3.10(*b*) since it is on top of the hood. Raising or lowering the sash however actuates a damper through a linkage arrangement so that the air is again proportioned between the hood interior and the direct exhaust. The hood shown in Figure 3.8 is of the proportional by-pass type. The sliding-glass-panel approach used in the Blickman hood to control face-opening area eliminates the proportional approach, so the hoods seen in Figure 3.9 are for use in the room by-pass situation and operate basically as shown in Figure 3.10(*a*).

3.5.3 Exhaust Ductwork and Plenums Just as the toe bone is connected to the foot bone and the foot bone is connected to the ankle bone, hoods must be connected to exhaust ductwork and that in turn to the air-filter plenums.

The size (internal cross-sectional area) of exhaust ducts is primarily determined by the quantity of air that must be handled, but it is also affected by the shape of the duct cross section. The main duct in many older chemistry buildings was square or rectangular and lined with tile. These square-angled configurations have the advantages of being easier to install and facilitate the making of necessary connections to the hoods and side ducts, but they present more resistance to air flow. The pattern is still used to a considerable extent but the use of round ducting is probably now more usual.

These are matters for the ventilation engineer as are the problems of proper duct supports and of design of plenums, filter enclosures and exhaust fans in order to reduce noise and vibration problems. The noise background in a laboratory through which large volumes of air are being moved can be obnoxious, simply because of the action-interaction of the air flow with the surfaces and movable pieces of hardware it encounters. The phenomenon is not on the plus side of comfort, and is a big argument, in addition to the economic and energy conservation aspects, for continuing efforts to develop better methods for reducing total air supply while still maintaining safety.

Some general rules relating to ductwork are the following: (1) The runout length should be as short as possible in order to reduce corrosion, condensation and future decontamination problems. (2) Horizontal runs provide more surface where condensation can occur and be retained than do vertical runs. (3) The exhaust fans in radioisotope facilities should be downstream from the HEPA filters so that the ductwork between the hood and the filters is kept under negative pressure to reduce the chance of leakage to the outside. (4) The ductwork should contain no flammable material, including any lining, and have a minimum of angled obstructions to the smooth flow of the air. (5) The design of the ducting as it leads into the filter plenum is particularly critical in eliminating noise and vibration problems. (6) The possible need for repairs and the inevitable need for eventual decommissioning should be kept in mind during design. (7) The duct construction material should resist corrosion, dissolution and any tendency to melt or catch fire under both normal operating and under accident conditions.[21c,30,38]

A number of duct construction materials have been utilized, with varying results. The radioisotope facilities considered in this book are heavily on the chemical side. This means that the ductwork must be able to handle fumes from acids and bases, corrosive gases, organic solvents and the results of an improbable but possible fire or explosion in the hood. All of this is obviously a large order.

ASHRAE lists as possibilities: (1) glazed tile, (2) cementitious materials (Transite), (3) galvanized iron, (4) stainless steels, (5) asphaltum-coated steels, (6) epoxy-coated steel and (7) plastic materials, the chemical resistance of which they list for eight types. This listing is reproduced in Table 3.5.

Hughes[37] advocates mild steel coated internally with a plastic compound, seven of which he lists. (Most of these appear to be British proprietary products.) Nixon and Chapman[21c] simply state that mild steel is inadequate as a duct construction material. Aside from chemical suitability, the product chosen must necessarily also be considered in terms of cost, of ease of fabrication and installation and consideration given to the problems of repair and of eventual decontamination and dismantlement.

Webster's dictionary has as one of the several definitions of plenum that of an enclosed space at a higher pressure than that of the outside. In ventilation

Table 3.5 Plastic Ducting Chemical Resistance[a, b]

Generic Type	Acids		Alkalies	
	Weak	Strong	Weak	Strong
Polyvinyl Chloride (PVC)	X	X	X	X
PVC-coated steel (P.V.S.)	X	X	X	X
Polyetheylene	X	VS	X	X
Fluorocarbon[c]	X	X	X	X
Phenolic (asbestos-fiber reinforced)	VS	VS	S	A
Polyester (glass-fiber reinforced)	X	S	S	A
Epoxy (glass-fiber reinforced)	X	S	X	S
Furan (glass-fiber reinforced)	X	X	X	S

Source: Adapted from Table 3, Chapter 15, *ASHRAE Handbook 1974*. Reference 30.

[a] X indicates no attack or generally insignificant; VS, very slight attack; S, slight attack; and A, attacked severely.

[b] PVC is severely attacked by aromatic or ketone solvents, inert to aliphatics. Polyester is rated as S towards organic solvents, all of the others listed as X.

[c] The VS rating for polyethylene and the phenolic compounds is restricted to oxidizing acids.

work the term is more generally used to indicate a broadening of the cross section of a duct (and perhaps a change in its geometry, e.g., from round to rectangular) as it approaches a nonduct structure such as the unit containing the HEPA filters. Broadening the cross section reduces the air velocity somewhat and produces some mixing, but the main purpose of a plenum is basically that of mating with the next equipment in the line.

Plenums are generally constructed of the same materials as used for the duct run-outs, although this may not always be true. Plenums should be designed to avoid collapse due to possible high negative pressures being generated in the units between the HEPA filters and the air inlet. The plenum wall thicknesses can be substantially reduced by the use of stiffeners.

3.5.4 Filtration Filtering of air is not unique to nuclear installations, being similar to numerous other applications throughout industry, but for obvious reasons the efficiency and dependability of the filters are particularly critical when radioactivity is involved. There accordingly has been much research, still continuing. The (AEC-ERDA) DOE has for a number of years acted as cosponsor (generally with the Harvard University Air Cleaning Laboratory and sometimes others) of large biennial conferences[31,42,43] on the subject. Many of the papers at these meetings are concerned with the problems particularly pertinent to nuclear reactor and fuel-reprocessing exhausts, but there is generally at least one complete session and other odds and ends among the general papers devoted to the question of filters and filtration. Another excellent source of information is the "Nuclear Air Cleaning Handbook" originating at Oak

Ridge,[44] prepared by a team headed by Burchsted, and subtitled, "Design, Construction and Testing of High-efficiency Air Cleaning Systems for Nuclear Application." Standards abound. The AEC and its successors[45] appear to generally follow the specifications established by the Department of Defense[46] and the Institute of Environmental Sciences[47] (formerly the American Association for Contamination Control). The Underwriter's Laboratories have several applicable standards,[48] and the specifications issued by a number of other organizations apply to the materials used in filter construction. The testing of filter efficiencies currently largely goes by ASHRAE Standard 52-68.[49] Filter manufacturers naturally test and certify their products, but in addition the government operates independent quality-assurance testing facilities at Oak Ridge and at the Hanford reservation in Washington state.[50] A purchaser can instruct the vendor to forward the filters to one of these locations where each unit is individually tested. This service is free to certain government agencies and their contractors, a charge (about 10% of the cost of the filter) is made to others.

Filtration is of course of little use in eliminating radioactivity if it is in the form of a pure gas. Such gases are relatively few in number, but of considerable importance in some situations. Usually the radioactivity carried in the ventilation exhaust stream will be in the form of aerosols or particulates or associated with such suspended materials (fumes, smokes, dusts, mists). The suspended particles can be large enough to be visible, as in the case of dust from an uranium ore grinding operation, but are more generally microscopic or submicroscopic in size, ranging down to a few hundreths of a micrometer in diameter. Efficient removal of these ultrasmall particulates from large volumes of fast-moving air is the major challenge to the filtration experts.

The following discussion will deal briefly with the other filter problems in a nuclear facility—the supply air and the question of prefilters—but the main emphasis will be on the HEPA final filter. Other alternatives are available for decontamination of the exhaust air, but the HEPA approach dominates. The Savannah River Plant in South Carolina has for years used deep beds of sand to filter their exhaust air with quite good results.[43a] The approach however consumes much space and is obviously best suited for a very large installation discarding corresponding enormous volumes of exhaust.

Liquid scrubbers and sprays, cloth bag filters, centrifugal separators, inertial and mechanical separators, and electrostatic and ultrasonic devices are other possibilities. Some of these approaches can be applied in special cases, but for all practical purposes the nuclear industry depends primarily on HEPA filtration as the final clean-up step before releasing the exhaust air to the external surroundings.

Filters for the supply air are primarily to remove atmospheric dust; partly to reduce the dirt load on the building exhaust filters, and partly for purely housekeeping reasons. A good air-supply filter can materially reduce janitorial costs and appreciably extend the period between major cleaning and repainting operations. Filters of 50%–65% ASHRAE Standard 52-68 efficiency are applicable. Louvers, moisture separators or both, as well as some sort of heating device in some regions for eliminating ice, should be provided in order to protect the filters from the weather. Clogging by grass clippings, leaves and wind-blown trash can be minimized by locating the air-intake opening above ground level.

Prefilters were mentioned in connection with Blickman-style hoods where panel-type filters are part of the unit. This arrangement is not of course a requirement and the prefilter can be anywhere in the ventilation run before the final filters, or even left out entirely. The Oak Ridge handbook[44] however shows an interesting pair of graphs taken from the text by White and Smith,[51] *High Efficiency Air Filtration*. The pressure drop across an unprotected HEPA filter in about 11 months will increase to the point where the unit should be replaced. The life of the same HEPA filter is prolonged to about 2½ years if prefilters, changed every 14 months, are used. HEPA filters vary in size and cost, but a full-sized unit runs well over $100, while prefilters cost only a fraction of that amount. A careful accounting analysis should therefore be made before deciding on elimination of prefilters.

For clean and new HEPA filters of any size operated at rated air flow, specifications call for a pressure drop of 1 in. of water (1 in. wg, or water gauge) across the unit. This pressure drop increases as the filter becomes dirty and most manufacturers recommend that a unit be changed at 2 in. wg. The tests by White and Smith described above were taken to 4 in wg, probably standard practice in most nuclear installations. HEPA filter specifications call for the unit to withstand up to 10 in. wg without damage or loss of efficiency, but working to such a high level would necessitate higher static-pressure fans, larger motors and heavier ductwork,[44] again raising the question of original investment versus operating costs.

Table 3.6, taken from the Oak Ridge handbook, shows the comparative efficiencies of the four general classes of filters as defined in ASHRAE 52-68 where the details of testing methods are also given. Group-I panel filters (viscous impingement filters) are shallow traylike assemblies of coarse fibers (glass, wool, vegetable or plastic) or of crimped metal mesh, enclosed in a steel or cardboard casing. The medium is coated with a tacky oil or adhesive to improve retention of trapped particles. In some types the medium may be replace-

Table 3.6 Comparative Filter Efficiencies

Group	Removal Efficiency For Various Particle Sizes (In Micrometers)			
	0.3	1.0	5.0	10.0
I	(0−2)%	(10−30)%	(40−70)%	(90−98)%
II	(10−40)%	(40−70)%	(85−95)%	(98−99)%
III	(45−85)%	(75−99)%	(99−99.9)%	99.9%
HEPA	≥99.97%	(99.99)%	100%	100%

Source: ORNL Nuclear Air Cleaning Handbook. Reference 44.

able or recleanable for re-use; others are manufactured as throwaways. Group-I filters are best used to trap fibrous and visible dusts, but are almost completely ineffective in removing particles smaller than 5 μ. Groups II and III are extended-medium (deep) dry-type units—the medium is pleated or formed into bags and housed in a box rather than a tray to give a large internal surface with a minimal frontal area. The medium is not coated. Replaceable and cleanable media and throwaway types again are available. Group-II filters are recommended for high-lint and -fiber loading applications. Group-III filters are more effective for removal of smaller particles, but the dust-holding capacity is generally lower.

HEPA filters are of five standard sizes, the airflow capacity varying from 25 to 1000 cfm or more. The units are boxlike in shape, the largest having face dimensions of 2 ft on a side and a depth of 11 ½ in. Special small filters in the shape of a stubby cylinder are also available for special applications such as in gloveboxes or in-cell containment boxes. The filter medium itself is usually woven of glass fibers or glass fibers and asbestos paper. The medium in most types is formed into a corrugated shape and then folded back and forth on itself so that it becomes self-supporting. In other versions the medium is folded over metal or plastic separators for support. Figure 3.11 shows the front face of a HEPA filter. Pleating of the medium of course makes much more surface available for filtration. The 2 × 2 ft (exterior face) units produced by different manufacturers have internal filtration areas ranging from 240 to 320 ft² and when clean vary in air volume handled per unit over an approximately 1000–1800-cfm range.

Gunn and Eaton[43] have reported on a grueling set of tests carried out on six HEPA filters obtained at random from American manufacturers. Some units came through better than others.

The effects of a sudden change in pressure on a filter that would result from the passage of a tornado is of obvious concern and a combined theoretical-

Figure 3.11 High-efficiency air filtration (HEPA) unit. (Courtesy Flanders Filters, Inc., Washington, N. C.)

experimental program to study the question has been established[43c] at Los Alamos Scientific Laboratory. More work needs to be done, but preliminary indications are that present-day HEPA filters would hold up to such a pressure change, particularly since the latter would probably be significantly attentuated because of the stack, ductwork and fans downstream from the filters themselves.

Another matter of concern is the behavior of the HEPA filter in case of fire. A fire in a hood or in the ductwork that generated a heavy smoke could cause plugging of the filter and result in overpressure of the hoods or gloveboxes and possibly force radioactivity back into the room. Because of the importance of this problem the government has built a full-scale test facility at Lawrence Livermore Laboratory.[43d] This unit allows testing of the effects of burning different types of combustibles, of varying duct configurations and placement of the prefilters and fire-retarding screens, the possibility of washing out the smoke particles by water sprays, etc. The use of water sprays was also studied earlier at Livermore[42a] from the point of view of protecting the filters from heat effects. The paper quotes the following numbers for the heat degradation of HEPA filters:

Temperature °F	Service life
750	Less than 10 min
325	Up to 2 hr
276	48 hr

First[42b] has also studied filter behavior up to 1000 °F. Filtration efficiency decreased somewhat in the 800-1000 °F range and in some cases the filter casing warped and put stress on the medium.

One rather unusual detriment to the effectiveness of HEPA filters is that of penetration by alpha-particle recoils. As an alpha is ejected from a nucleus the action-reaction causes the displacement of surrounding atoms or even aggregates of atoms. The recoils can of course occur in any direction, but the final effect of a heavy loading of alpha emitters on a filter is that the activity works its way into the filter medium and can eventually cause enough structural damage to cause an observable drop in the filter efficiency. The same action presumably occurs with beta emitters but here the recoils are very much smaller because of the lighter mass of the ejected particle. The effect of alpha recoil on filters was studied by McDowell, Seeley and Ryan.[43c] The effect of high levels of gamma irradiation was examined by Cheever and co-workers.[42c] Filters were irradiated at different exposure levels up to 1.5×10^9 R with little effect on efficiency.

The HEPA filter performance specification is: ". . . shall have a maximum penetration of 0.03% (i.e., a minimum of 99.97% efficiency) when tested with monodisperse dioctylphthalate (DOP) smoke having a light-scattering mean droplet diameter of 0.3 micron when operated at rated airflow capacity and at

20% rated air capacity."[47] Many commercially available units have efficiencies approaching 99.99%, that is, one 0.3-μ particle in 10,000 will penetrate the filter. Even this is too high in some cases, so filters can be placed in series; a common practice in plutonium facilities, nuclear plants and hot laboratory operations. The filters following the first will not have equal efficiency since the particle-size distribution is changed to favor the harder-to-remove smaller particles at each stage. Gonzales *et al.*[31a] studied the effectiveness of three HEPA filters in series against penetration by $^{238}PuO_2$ aerosols. The array gave decontamination factors (D. F.'s*) in the 10^{10}–10^{12} range.

The most commonly used agent for testing filter efficiency is dioctyl-phthalate (DOP). This material can be volatilized under controlled conditions to form a cloud of particles of known size distribution. A fixed amount of DOP smoke is released into the air flow into the filter and the number of particles of a given size penetrating the medium determined by light-reflection techniques downstream. Sodium chloride can be similarly utilized, and piezoelectric, laser and other approaches to filter testing are being developed. Part of the problem is purely analytical because of the difficulty of detecting and characterizing the small number of particles making it through the filter.

HEPA filters in a radioisotope laboratory will very seldom become a shielding problem because of the loading of radioactive particulates, although this can become a headache in the filter room of fuel-reprocessing plants and other industrial nuclear facilities. Removal of a spent filter and installation of its replacement can still be a chore because of the chance of spreading contamination. "Bagging" of materials into and out of a contaminated interior volume is a technique of long standing in the nuclear world and is frequently used in making filter changes. Figure 3.12 shows the general principles of bagging, although there are changes in detail depending on the size of the items to be moved back and forth and the structure of the contaminated enclosure. A more complete diagram of the rather sizeable effort needed to replace a highly contaminated full-size HEPA filter is given in the Oak Ridge handbook.

The bagging process is described by its name. The idea is to remove material from the inside of a highly contaminated enclosure, to get that material into a sealed plastic bag in the fashion prevalent for foods in the supermarket, to replace the removed item with a clean unit if needed and to conduct the entire operation without allowing any of the radioactivity to contaminate the operators or the outside. There is a constant effort to improve the devices to make all this possible, and double or even triple bagging is not at all unusual. The basic pattern is somewhat as follows. A metal ring or frame is built onto the enclosure around the bagging opening at the time of construction. In its most primitive form, this ring or frame will have two circumferential channels into which clamping metal rings can be fitted to make an air-tight closure of the open end of a plastic bag. Initially the bag will be clamped into place using the inner (closest to enclosure wall) channel. When the time comes to move an item out of the enclosure, the material is pushed out into the bag which is then

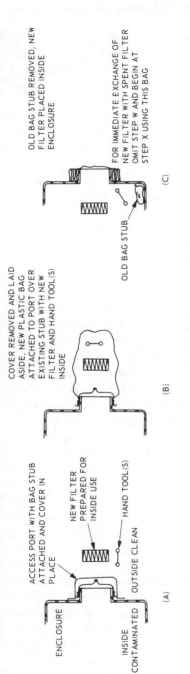

ENCLOSURE

ACCESS PORT WITH BAG STUB ATTACHED AND COVER IN PLACE

NEW FILTER PREPARED FOR INSIDE USE

HAND TOOL(S)

INSIDE CONTAMINATED   OUTSIDE CLEAN

(A)

COVER REMOVED AND LAID ASIDE, NEW PLASTIC BAG ATTACHED TO PORT OVER EXISTING STUB WITH NEW FILTER AND HAND TOOL(S) INSIDE

(B)

OLD BAG STUB REMOVED, NEW FILTER PLACED INSIDE ENCLOSURE

FOR IMMEDIATE EXCHANGE OF NEW FILTER WITH SPENT FILTER OMIT STEP W AND BEGIN AT STEP X USING THIS BAG

OLD BAG STUB

(C)

NEW FILTER BAGGED INTO CONTAMINATED ENCLOSURE

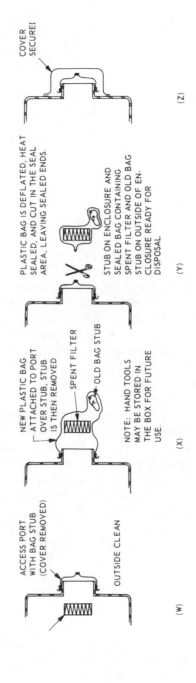

ACCESS PORT WITH BAG STUB (COVER REMOVED)

OUTSIDE CLEAN

(W)

NEW PLASTIC BAG ATTACHED TO PORT OVER STUB, STUB IS THEN REMOVED

SPENT FILTER

OLD BAG STUB

NOTE: HAND TOOLS MAY BE STORED IN THE BOX FOR FUTURE USE.

(X)

PLASTIC BAG IS DEFLATED, HEAT SEALED, AND CUT IN THE SEAL AREA, LEAVING SEALED ENDS.

STUB ON ENCLOSURE AND SEALED BAG CONTAINING SPENT FILTER AND OLD BAG STUB ON OUTSIDE OF ENCLOSURE READY FOR DISPOSAL

(Y)

COVER SECURED

(Z)

SPENT FILTER BAGGED OUT OF CONTAMINATED ENCLOSURE

Figure 3.12 Bagging-in and -out of a containment enclosure. (From Ref. 44, ORNL drawing 69-8774.)

67

heat sealed at the top next to the box. A careful cut is then made across the seal, leaving the hot material sealed in its bag and a sealed stub hanging on the enclosure ring. The clamp on this stub is then moved to the outer channel and a new bag containing any items for transfer into the enclosure fastened over the stub and clamped onto the inner channel. The old stub can then be released to be pushed into the enclosure or into the new bag for disposal during the next bagging operation. It is then good practice to fold the new bag into the ring and to close the opening with a coverplate until the next time around in order to avoid accidental removal of the bag.

After a number of such operations in the simple arrangement described, some activity eventually creeps out into the two-channel interface and immediately starts showing up everywhere else. Various more sophisticated approaches have accordingly been devised; involving more clamping channels, ingenious clamping devices and additional heat seals, but the basics remain pretty much as indicated. Bagging out of dirty HEPA filters can become a time-consuming and at least a mildly hazardous business in a nuclear plant, so the filter rooms in such establishments have to be designed with this in mind, in addition to provisions for *in situ* testing. Figure 3.13 shows a filter containment unit having many of the needed fixtures. DOP testing inlets are shown and the bagging connections are presumably in back of the bulkhead door.

The radioactive gases that have received the most attention from the point of view of removal from air streams are quite naturally those that are produced directly in fission or as a result of other nuclear reactions in a reactor. These gases include iodine, krypton, xenon, tritium and, somewhat to everyone's surprise, it now appears[31b] carbon-14 dioxide. At the Fourteenth Air Cleaning Conference[43] seventeen papers were presented dealing with some aspect of iodine clean up, eight on the rare gases (Kr, Xe), two on tritium and one on ^{14}C. These effluent gases from the nuclear fuel cycle are of course considered to be part of the radioactive-waste-disposal problem, so at the end of the conference a panel discussion[43f] was held under the auspices of what was then the ERDA Nuclear Fuel Cycle and Production Division in order to review the state of the art of trapping radioactive gases from the engineer's point of view. Much of the information in the following paragraphs is taken from that report.

It is thought that iodine is released from the nuclear fuel cycle as a mixture of the elemental form, methyl iodide, and hypoiodous acid (HIO).[52a] Absorption of the iodine on activated-charcoal beds is the most commonly used current removal method, but is not entirely satisfactory since some beds for reasons not understood are readily poisoned and end up with uneconomically short service lives. Caustic scrubbing of the air stream has been tried but apparently now abandoned. Scrubbing with mercuric nitrate or boric acid–nitric acid solutions is apparently more effective, but of course this approach generates a liquid-waste disposal problem to replace the original gas-elimination difficulty. Silver-coated absorbents (silica gel, alumina, amorphous silicic acid and zeolites) are all receiving attention and a silver zeolite will be part of the Barnwell

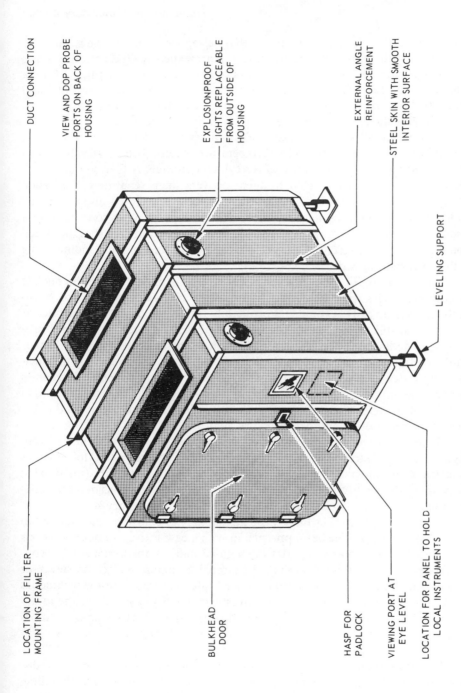

Figure 3.13 Steel HEPA-filter enclosure. (From Ref. 44, ORNL Drawing 69-8736R.)

DUCT CONNECTION

VIEW AND DOP PROBE PORTS ON BACK OF HOUSING

EXPLOSIONPROOF LIGHTS REPLACEABLE FROM OUTSIDE OF HOUSING

EXTERNAL ANGLE REINFORCEMENT

STEEL SKIN WITH SMOOTH INTERIOR SURFACE

LEVELING SUPPORT

LOCATION OF FILTER MOUNTING FRAME

BULKHEAD DOOR

HASP FOR PADLOCK

VIEWING PORT AT EYE LEVEL

LOCATION FOR PANEL TO HOLD LOCAL INSTRUMENTS

Nuclear Fuel Plant's exhaust system if that beleaguered facility is ever allowed to proceed with reactor fuel reprocessing. An inexpensive iodination box has been described[53] for radiochemical laboratory use. The box is a cube, 1 ft on a side. The unit has its own blower which draws air through two portholes and exhausts through charcoal beds. The box can be either placed in a hood or on a bench with the exhaust delivered to a hood by flexible hose.

The rare gases (Xe and Kr) are held to a certain extent on charcoal beds, particularly if these are kept at low temperature. After a bed is saturated it can be warmed and the gases drawn off for cylinder storage. A German method[43g] solidifies the Xe and Kr at 80 K, then recovers them separately, the xenon being used industrially after the short-lived species have died away. An Oak Ridge process[43h] absorbs the krypton-85 (the isotope of most concern) into a fluocarbon solvent (CCl_2F_2) with over 99.9% efficiency and furnishes the added bonus of removing essentially all of the elemental and organic iodine along with the carbon dioxide.

Tritium (^3H or T) removal from a gas stream is difficult since the nuclide rapidly exchanges with the normal hydrogen of any water present and essentially all ends up as HTO or T_2O. Fortunately tritium production in the nuclear fuel cycle is still at rates markedly lower than that produced by natural processes.[52b] Methods for keeping the tritium out of the exhaust stream by "head-end" treatment are being studied, that is, trapping the gas during the fuel-dissolving step at the reprocessing plant. Various isotope-enrichment schemes are also being investigated in order to produce enriched fractions more readily handled as waste. References giving handling procedures for tritium and information on monitoring and detection are available.[54,55]

There has not as yet been much reported investigation of ^{14}C removal from high-volume gas streams, but the problem is not expected to be too difficult because of the chemical reactivity of the probable product forms (the dioxide or possibly the monoxide or methane). There of course is a considerable background of experience in handling the isotope on the laboratory scale.[56,57]

Radioactive gases other than those named may of course be encountered in radiochemical work. The best approach in such a case is to carry out operations in a closed system so as to restrict the volume that eventually must be decontaminated. Gas scrubbing, activated-charcoal columns, molecular sieves or other solid absorbents can be used to concentrate the activity; or the techniques used in the chemical industry can be investigated.[52a,58] In some cases the radioactive half-life will be short enough to allow transfer of the gas into a temporary container for storage until the activity has died away.

3.5.5 Gloveboxes There has been one notable exception in the above discussion of ventilation matters, and that is the use of gloveboxes in handling radioactivity. This is a deliberate omission because of the recent very thorough review of glovebox techniques by C. J. Barton[59] of Oak Ridge, an update of his earlier chapter[60] on the same subject.

Gloveboxes are in a sense totally enclosed hoods, but there are differences. The air needed for ventilation is brought in through a filtered intake, and in many boxes the first HEPA filtration of the exhaust takes place immediately at the box. Provisions such as an air-lock entry port on one face of the glovebox must be made for transferring materials in and out. Protection of the operator depends on physical barriers (the walls of the box and the gloves) so there is not the same need as in a hood for high-velocity air flows as a safety measure. Ventilation of a glovebox thus becomes primarily a matter of heat and fume removal and air flows can accordingly be very much reduced. Whereas a wide-open radiochemistry hood might require 1500 cfm or more of supply air, a glovebox of the same interior volume could well get by with less than 10% of that amount. On the other hand, since the need for glovebox use implies the handling of larger quantities of radioactivity, particularly alpha emitters, the exhaust-air cleanup problem is exacerbated.

3.6   LABORATORY SERVICES (UTILITIES)

The radioisotope facility will in general require the same utilities as any other type of modern laboratory: cold, hot, distilled and perhaps chilled water; natural gas; compressed air; electrical power; perhaps a few special "house" gases (nitrogen, oxygen); and perhaps steam. The radioisotope laboratory probably differs most from at least some other types in having to establish a separate system for disposal of routinely-produced, possibly hazardous liquid wastes. One utility that is usually missing when radioactivity is being handled is a house vacuum system. The accidental contamination of such a system at one location would make the whole of dubious value throughout unless the contaminating nuclide was of short enough half-life to quickly die away.

3.6.1   Distribution to the Laboratories   The separate services will be briefly discussed later, but first there is the problem of getting them to the point of use, a question closely related to facility layout as covered in Section 3.4. In addition to getting things in, there are a few that have to taken out such as sink wastes and steam condensates. There are also other supplementary services such as telephone and computer connection lines, public address systems, emergency alarms, perhaps automatic sprinklers for fire protection, etc., whose distribution may or may not be able to take advantage of the passageways needed for the plumbing, electrical and other service hardware or those carrying the ductwork and other equipment associated with the HVAC system.

Most of the main supply lines for the utilities will come into the building at, near or below ground height so the service lines will usually first have to rise vertically to the laboratory level then be teed-off at right angles horizontally to be distributed to the individual modules or rooms, frequently with one or more right-angled changes in between. Unless the vertical portion is brought up directly to each module, the horizontal component will probably be in the form of a stylized tree—a central trunk (down the ceiling of a central corridor as an example) with branches off at right angles at each module position, Pipe sizing of course decreases in each section of the system from the main supply line to the point of use and similarly increases for the waste lines going back in the reverse direction. In a one-story structure with no service floor (basement) the vertical component is largely eliminated, but a method still has to be chosen for the interior horizontal distribution. If such a one-story building does have a service floor the feeder mains can be carried along its ceiling and fed upwards through holes in the ceiling into the modules above. Whenever this pattern is followed the penetration holes should be carefully sealed around the piping (or a dike installed) to avoid leakage into the lower area if a minor flooding incident occurs on the upper floor. Such penetrations must also be examined from the point of view of their adding to the air-ventilation balance problem or in possibly aiding in the spread of fire in case of a conflagration.

The plumbing bill is one of the big ones in laboratory building construction so considerable care should be taken to select a distribution scheme that will minimize unnecessarily long pipe runs or complicated connections. Valving should however to be adequate enough to permit one module to be closed off for repair work without disrupting service elsewhere, and generous instrumentation for determining pressures, temperatures and other relevant control parameters is bound to be a good investment. There will be the usual number of built-in contradictions. Exposed piping and conduits should be kept to a minimum in a radioisotope laboratory in order to reduce the chore of keeping the area dust free. On the other hand, while burying the service lines in or in back of the furniture simplifies housekeeping and improves the aesthetics, it complicates the problems of maintenance and repair and increases the difficulty of making a quick change in the basic pattern, that is, flexibility is decreased.

The problem of bringing service-supply feeder mains through the vertical components of a tall building has a number of solutions, usually based on utility shafts which may be relatively small if one goes to each individual module position or quite large if only a few shafts are used and ductwork for the HVAC system is also placed in the same passageways. More details are given in the articles by Ruys,[10a] Mellon,[9] and Solomon.[6b] Other relevant references are by Hughes and Cullingworth[61] and the earlier NAS-NRC book[5] on laboratory design.

3.6.2 Distribution within the Laboratory

Utility piping and conduits can be brought into the laboratory in various ways. If the furniture is along the side walls and/or formed as a peninsula out from one of the end walls, the piping can be brought in horizontally below or just above desk level or can enter the room at ceiling height in a modular pattern to be led down to the desk outlets. The desk-height approach obviously will not work for free-standing units such as a vacuum frame hood or island-type desk arrangements without impeding personnel movement around the unit. The down-feed method is obviously better suited for such situations. As an alternative the services could be brought to such free-standing units in floor trenches, but the associated floor excavations would require cover plates with the implication of trench contamination in the event of a radioactive spill.

The trench approach can be carried even further by utilizing a false floor. The services are brought in under a raised flooring that serves as the working area. This requires cover panels in the false floor to allow access to the piping and thus again is not suited for radioisotope laboratory design.

The method most commonly used is to bring the service lines in at or near desk height. The horizontal pipe runs for either peninsular or side desk arrangement can be supported within the furniture itself or can be carried along the back of the desk tops, either exposed or enclosed in some type of curb box. The piping can also be hung on or buried in the walls for the side desks, or supported on a free-standing pipe rack between the furniture and the wall. As an extension of this last approach, the pipe rack can be made sturdy enough to support concrete blocks built up to the ceiling to form the laboratory wall. The exposed piping in the lower portion is normally covered by the furniture or removable metal panels. The utility outlets are distributed in modular fashion in a box curb along the back of the desks. This curb is supported independently of the furniture units so these can be moved out fairly readily if access must be gained to the piping.

The lines coming down from the ceiling in a down-feed system to the outlets on desks in the peninsular or island arrangements should be braced in some manner. They are usually left exposed. The down-drop lines to the side desks can again be buried in or hung on the walls or independently supported. The incoming horizontal piping along the ceiling can be left exposed or preferably covered by a false ceiling to minimize the accumulation of possibly contaminated dust on hard-to-clean surfaces. The down-feed system has the advantages of making most of the piping more accessible and of simplifying the use of free-standing equipment and furniture. The overall building however becomes taller since each laboratory must be of greater height in order to accomodate the false ceiling.

As indicated earlier in the chapter, the British have had a continuing study[17] under way for some time aiming to improve laboratory flexiblity. Their Department of Education and Science established a Laboratories Investigation Unit in the late 1960s, the members of which have developed a system now being marketed under the name of "Metriscope" which has already been established in a number of English buildings, chiefly at universities. The design of the utility distribution is essentially down feed. A grid of service outlets is established at the ceiling level over an entire laboratory area (inner walls are movable partitions), and a matching grid of drainage outlets in the floor. The upper grid is covered by a false ceiling with boxed-in openings at the utility take-off points. The drainage take-offs in the floor are covered with metal plates.

Benches, desks and hoods are all freely movable, each being equipped with a set of tubes extending almost to the false-ceiling height. These tubes carry the utility lines for that particular unit. When one of these is moved to a new position, connections are made to the nearest service point by flexible tubing or electrical cable. Sinks are also movable, and again, if placed in a new location are connected to the nearest drainage take off with plastic hose. The system generates some drooping hoses and electrical cable just below the ceiling, but this is minimized by the height of the rigid tubes carrying the services for most of the distance down to the furniture unit.

Photographs of some of the installations using the Metriscope system show laboratories that are quite acceptable from the aesthetics point of view. The approach is most interesting and certainly has considerable promise for regaining some of the flexibility in making internal rearrangements that has been largely lost in more conventional laboratory design systems. The big question in applying the Metriscope approach to a radioisotope facility would be that of adequate handling of the air-supply and -exhaust requirements.

3.6.3 Laboratory Walls and Partitions It should ideally be possible to change room sizes practically overnight by the use of movable partitions. The Metriscope system described above implies that this can be done, but the reports available do not give much detail on the method of accomplishment. Readily movable room dividers are used extensively in modern office buildings, but the problem becomes more difficult for laboratories where general safety concerns and building code requirements for fire- and soundproofing necessarily are more stringent.

Movable partition walls are made much more conceivable in the Metriscope concept because the major portion of the horizontal utility piping is tucked out of the way in the ceiling or floor and the vertical components are chiefly on the furniture itself rather than in, on, or in front of the walls. (Ductwork could be a complication.) Such is usually not true in more conventional systems. A substantial amount of the service hardware in these must almost invariably be dismantled and be rebuilt elsewhere if a wall position is changed—a frustratingly time-consuming and expensive operation that considerably diminishes the advantages of movable partitions since they turn out to be not all that immediately movable.

Pipe hangers, piping curbs, wall cabinets, etc., should not be hung on movable partitions because the necessary drill holes will probably be in the wrong place for the next application of the partition panel. (The Metriscope system gets around this by having channels of the Unistrut type on each panel. Equipment is hung from this channel, but nothing is ever permanently fastened to the wall.) If movable partitions are used, they should be highly standardized in a minimum number of designs and assurance should be obtained that off-the-shelf replacement units will be available from the supplier in the future. Knocking down a plastered concrete-block wall, while admittedly a dusty and noisy operation, should be considered as a viable alternate to movable partitions unless it is anticipated that there will be a long series of room changes after occupancy.

3.6.4 Specific Utilities

The hot- and cold-water supply to the laboratory sinks should be a system completely independent of the one used for the domestic water; that used for drinking, sanitary and janitorial purposes elsewhere in the building. This is of course to eliminate any possible contamination of the potable supply by accidental backup from the laboratory system. While the domestic water waste can be disposed of directly to the building sewage system, the laboratory drains should empty into large retention (hold-up) tanks so that the contents can be checked for radioactivity before being released (Figure 3.5). If the activity level is above tolerance level the tank contents can be pumped into a tank truck and taken to a treatment facility for cleanup. The tank otherwise can be drained to the regular sewage system, by far the usual situation.

Such an elaborate arrangement may of course not be feasible for a small laboratory. Extreme care must then be taken to be certain that radioactive liquids are collected in as concentrated a form as possible for transfer out through the administratively established waste-disposal mechanisms and not poured down the sinks. Sink outlets should be frequently monitored to determine that the rules are being followed.

Retention tanks generally come in pairs so that while one is being filled the other is being checked for activity content or being drained. A problem obviously arises if the tank being filled reaches capacity before its partner is empty which can happen if there are relatively constant high-volume flows of waste from the laboratories. Aspirators are a convenient source of vacuum but are also such high-volume waste producers in addition to being a possible source of tank contamination. Aspirators should accordingly be used sparingly or not at all and vacuum obtained by means of a trap-protected pump when needed. A frequently even larger source of drain waste can be water needed for cooling laboratory equipment since in many cases this has to be a continuing long-term operation. A separate chilled water line with the effluent going directly to the sewer can solve this problem if it is clear that there is no chance of leakage that would contaminate the outflow.

Distilled water for the laboratories in a large building has in the past usually been furnished from a central still and distributed in block tin, tin-lined brass, aluminum, stainless steel, or plastic piping. This approach deserves examination. Commercial suppliers have in recent years developed a variety of small and compact units for producing a high grade of deionized water by using combinations of mixed-bed resins, charcoal beds and membrane filters. The use of units of this type in each laboratory or for supplying a small group of laboratories could in some cases reduce original installation costs and provide water of equal or better quality than a central system with its extensive lengths of pipe runs where metallic contamination might occur. The point-of-use approach would also eliminate the problem of having all parts of the building affected by a breakdown in a central system. The use of many small units might on the other hand generate too much added maintenance so the question should be examined carefully.

It may be an illusion but the author has the impression that natural gas as a heat source in laboratory operations has been almost universally replaced by hot plates and other electrical heating devices and that Professor Bunsen's burner has lost much ground. There is still however at least one remaining important application for natural gas and that is in glassblowing. The service is therefore a necessity although conceivably could be brought only to a selected bench in each module where all glass working would be done. This however might generate problems when glass vacuum lines were being constructed in hoods in other parts of the room and there are undoubtedly still other occasional uses for gas. Modular distribution of natural gas outlets throughout the laboratory still appears to be justified.

The need for separate lines for distribution of special house gases will vary according to the type of research carried out by different organizations; the ready availability of smothering gases such as nitrogen, helium or argon in laboratories handling pyrophoric materials being an example. Nitrogen and oxygen (from exterior liquid-gas tanks rather than cylinder banks) were distributed to all of the Argonne Wing M laboratories. The experience has been that the nitrogen is quite frequently used, the oxygen essentially only for glassblowing. It can be concluded that only one or two oxygen outlets per laboratory would have been adequate or even that the O_2 system could have been entirely eliminated and reliance placed on cylinder gas as a source when oxygen is needed. The cylinder approach also offers flexibility if there are uncertainities as to the types and quantities of special gases that will be needed.

Compressed air is still a desirable utility to have available although it is used relatively infrequently, often being replaced by the house or cylinder nitrogen as a more inert and cleaner substitute. Steam as a distributed service probably receives even less use than compressed air in a radioisotope laboratory although there are obviously other facilities where this would not be true. Since both supply and condensate return lines for steam service outlets will be needed, the extra cost of supplying the utility on a distributed basis is usually not justified in a radioactivity laboratory unless some highly specialized operations are expected. The story might be quite different in semiworks or pilot plant facilities.

The question of electrical-power needs for an average radioisotope laboratory also requires a hard look if the budget is tight. Almost all normal requirements can be satisfied by a generous distribution of 110-V, single-phase outlets so automatic inclusion of 208-V, three-phase power at every bench position may be an expensive luxury. Sources of such higher power should be brought up to each laboratory module and counting room so that the service can be carried into the room later if really needed. Wholesale distribution of 208-V outlets along the benches will however probably be a wasted expenditure of money in most cases.

Much of the material in the last few pages has been personal opinion with little useful offered in terms of details of plumbing and electrical installations, pipe sizing, materials of construction, and so on. Some such nuts-and-bolts information can be obtained from a number of the previously cited references.[5,6,10a,15,30] The McGraw-Hill Publishing Company also issues *Sweet's Catalogue File*.[62] This many-volumed compendium is a collection of catalogues furnished by hardware suppliers of all sorts of building and laboratory construction materials and equipment, kept up to date as expeditiously as possible. Many of these catalogues contain much valuable data.

3.7 LABORATORY DETAILS

Design details in the radioisotope laboratory are strongly influenced by the possible occurrence of contamination and the consequent need for cleanup. Horizontal surfaces and exposed piping that could serve as dust catchers are kept to a minimum and open cracks and inaccessible right-angled corners where activity might lodge are avoided as much as possible. Construction materials should have impervious, chemically resistant and easily cleanable surfaces, particularly for the furniture, hoods and floors.

One other generality has already been mentioned in the discussion of module sizes, that of having the aisles between the furniture wider than normal. Another general feature is that each laboratory should have at least two exits so that the occupants are not trapped if a fire or similar accident occurs between them and the usual room entrance. This last requirement is of course not unique to the radiochemistry laboratory, nor is the need for evaluation for fire resistance of all of the materials of construction. The National Fire Protection Association's [26b] NFPA Code 45, "Fire Protection for Laboratories Using Chemicals," discusses such matters in detail.

3.7.1 Floors, Walls, Ceilings
Wood as a material for the laboratory floor has some drawbacks—it is combustible, its use in construction leaves cracks, and it is porous and will soak up a radioactive spill. The basic floor in most modern buildings will be concrete, but bare concrete (rarely seen in any type of laboratory these days) in addition to being tiring to stand upon for a good part of the day is also porous enough to generate a difficult decontamination problem if a spill occurs. Sealed and painted concrete is somewhat better but the paint will eventually wear away. The obvious, and usual, answer is to chose a hard-surfaced, long-wearing, chemically resistant and attractive floor covering having some resiliency in order to reduce the fatigue factor. A variety of suitable materials is readily available, although the range is somewhat reduced for a radioisotope laboratory since some of the more rigid materials, such as those based on asphalt, can be used only in the form of tiles. Tile is usually the form of choice for flooring in office and industrial buildings since if there is damage in a restricted area the tiles can be rather simply removed and replaced, whereas such a repair is more difficult with the sheeting types of flooring. Tile in a radioisotope laboratory however means a number of cracks. This might be acceptable in a tracer-level operation if the tiles are kept waxed to fill in the

cracks, but even here sheets in the maximum maneuverable width are probably preferable. Hughes[38] in Britain recommends a high polyvinyl chloride (PVC) plastic as a floor covering. This might be questioned since in a fire PVC may decompose to yield highly corrosive chlorine gases.

DeWahl[6c] in the second NAS-NRC book on laboratory planning listed the manufactured floor coverings in terms of increasing cost: asphalt, vinyl-asbestos, linoleum, vinyl, and rubber based; probably still a valid ordering. Whatever the choice, coving should be installed at the walls, furniture and other places where the floor covering meets a vertical obstruction. This coving should be of a shape to facilitate cleaning and be sealed to eliminate cracks with particular attention paid to the corners.

For either fixed or movable walls the surfaces should easily cleanable and nonporous. Plaster over a concrete-block wall works out quite well if the plaster is coated with an appropriately resistant paint giving a hard and smooth surface. Similar paints should also be specified for any movable partitions and carefully checked for ability to resist cracking and chipping if the partitions are to be frequently moved.

Realistically, walls are very much less likely to become a decontamination problem than are floors unless there is a major incident, and ceilings even less so. With so many hard surfaces called for in the lower regions there would seem to be some justification for somewhat softer materials in a suspended ceiling for purely acoustical reasons. The ceiling material under surfaces should nevertheless be easily cleanable. Most suspended ceilings are in the form of large tiles which can be replaced relatively inexpensively, so can be considered as being expendable if the decontamination problem becomes that desperate. The worst arrangement in such a contingency would be exposed piping running across the top of the room with no intervening false ceiling buffer.

3.7.2 Furniture Placement The hood or hoods should be placed where there is a minimum chance of air disturbance. There are two schools of thought with respect to duct runouts. Keeping them as short as possible simplifies clean-up if they become contaminated, but in case of a fire in a hood one would prefer as much distance as possible between the blaze and the HEPA filters. Each hood should be next to a bench or table on at least one side for convenience in making transfers in and out.

Drains from laboratory sinks definitely should be kept as short as possible because of the potential contamination problem. The drains should be readily accessible for the same reason and thus should not be located in back of furniture or buried in walls. (Cup sink drains will probably have to be less accessible.) In the utilities distribution pattern where the service shaft or corridor is immediately adjacent to a laboratory wall the sink should be located at that wall. There should be a minimum of one main sink per module in most cases.

The need for cup sinks on a modular basis along the laboratory benches should be carefully examined since they require both supply and drain lines and other hardware. Experience has shown that they are infrequently used in most inorganic and physical chemistry programs.

3.7.3 Furniture Decontaminability as well as chemical resistance has to be considered in specifying sink construction material. The usual choice is stainless steel, but a high Ni-Cr formulation should be chosen if much hydrochloric acid is to be used. There are quite acceptable alternatives such as molded sinks of modified epoxy resin and resin-impregnated sandstone units.[63] The resin-based formulations can of course be molded so as to incorporate the drain board and basin in a single piece without cracks. These sinks are somewhat more resilient than steel units and thus a bit easier on glassware.

A knee-operated emergency eye-washing sink is a desirable feature in any chemical laboratory, preferably separate from the dish-washing sink. If space limitations are such that this is not possible a separate water line to the main sink can be furnished, connected to available[64] portable eye-washing devices by flexible hose. While this alternative is not as desirable, some measure of emergency protection is furnished.

A completely modular arrangement in a laboratory implies that the sink, hood and furniture units are all of the same length. For example, if the chosen dimension is 4 ft for all items, six of them can be placed along each wall of a 24-ft module and all locations are completely interchangeable. Matters seldom work out that neatly since there are support columns and other structural parts of the building, emergency doors, special equipment of odd size, etc., that must go somewhere, so the ideal pattern is difficult to maintain. An effort should be made however, although the concept is probably less important for the sinks and hoods since they are not as frequently moved. The slack may have to be taken up by a custom-made or improvised piece of special furniture if any of the other units are of odd size. One modular position adjacent to to the room entrance should be reserved for a hand-and-foot counter in laboratories handling high levels of activity as will be discussed in the chapter on operations.

Laboratory furniture in industrial and governmental laboratories is now predominantly of metal rather than wood, although this is probably still not always the case in older academic facilities. Wood of course is less desirable in a radioisotope laboratory because of the difficulty of decontamination and the presence of cracks which may become more pronounced with warping and age. Early steel furniture also had its drawbacks, chiefly the tendencey to corrode, but there has been a steady development of excellent chemically resistant and impervious surface treatments. What is more, these surface coatings are now available in a variety of colors so the laboratory designer can try his/her hand at interior design to avoid the drab institutional look.

A large number of reputable suppliers of laboratory furniture is listed in the buying guides cited elsewhere, each of whose catalogues present a wide variety of units designed to be suitable for every type of laboratory operation. The first generality to be made with respect to radioactivity handling is the obvious one that the units chosen be such as to provide the minimum chance of accumulating contamination and the maximum ease of decontamination. Sturdiness is the other important criterion since the unit may be called upon to support a detector shield or other heavy equipment. ERDAM, Appendix 6301[15] states that most laboratory furniture units will support 300-500 lb per leg corner.

Potential furniture suppliers should be requested to quote exact ratings for their products.

Ruys[10a] lists stone, impregnated stone, resin, cement-asbestos, laminate plastic, laminated hardwood, impregnated wool fiber, stainless steel, and glass as materials for fabricating laboratory bench tops. Hughes and Cullingworth[61] reporting on British practice add formica bonded to plywood, polyvinylchloride sheet, FMB grade steel, glassfiber-reinforced resin, polypropylene, melamine and polyurethane lacquers on hardwood, and Corning's "Labtop," a lithium aluminum silicate glass ceramic, to the list. Some of these materials, particularly the woods, are not suitable for work with radioactivity. The material chosen should have the predictable characteristics—a smooth, impervious, readily cleaned surface of good chemical resistance, preferably of some resiliency and of such a nature that sealing materials of equal quality are available for filling in of cracks during construction. The top is frequently supported independently of the furniture units, which means that these latter can be pulled out and interchanged more readily. If the tops are part of the unit, seals will have to be made when two tops abut against each other which means a cutting operation if there is a later rearrangement of furniture. Separate support of the top also allows the use of crack-free lengths sufficient to cover several module positions. Many suppliers furnish tops having a slightly raised section at the outer edge on the operator's side to confine spilled liquids. The flat working surface may also be carried up in unbroken fashion to a service curb at the back, depending on the fabrication material. The cracks left where any bench top abuts onto any other surface should be sealed and the bead of the seal ground smooth.

The British Laboratory Investigation Unit study descibed in Section 3.6.2 included some careful analyses of the locations wherein the ever-acquisitive scientist stores materials against some possible future need. Under-the-bench cabinets for such storage came out rather badly—these become the repositories of the heavier junk and are difficult of access at best. Under-the-bench drawers obtained a better rating. The lower levels of over-the-bench cupboards or shelving are generally utilized for more frequently used items, but some of these are in the trivia category such as over-large stocks of ice-cream cartons, polyethylene bags and such. The upper shelves are almost hopeless of access to anyone not of basketball player height and thus tend again to become loaded with items that sit in the same place for decades. Ruys[10a] makes the same point with respect to the higher shelves.

There however does not appear to be any consensus as to the ideal way to handle laboratory storage problems. The obvious answer in a large organization is a communal instrument depository into which items would be transferred after their original utilization and then made available for use by others. This idea is appealing on paper, but difficult to bring about. Many scientists would prefer giving up their children before releasing a piece of equipment that might possibly be used in their own program sometime in the vague future. In practical terms, storage of such items tends to be on the back of the working benches or in the hoods (very dirty glassware) in the radioisotope laboratory. The heavier equipment after use tends towards the service floor or fan loft where it remains for years or until it has completely decayed away.

The author has very few constructive comments to make on this perennial problem and can quote no authority having a satisfactory answer. An annual inspection by top authorities and subsequent forced house cleaning seems to help.

3.7.4 Lighting Stein[18] in his book points out that lighting-level specifications in this country for offices and schools doubled or even trebled between the 1950s and 1970, ". . . as if we had suddenly gone blind as a result of some national disease . . .," as one reviewer commented. This pattern has undoubtedly also held true in laboratory design over the same period. Fifty foot candles or even less at the bench level was once considered adequate but, at least until very recently, standards have steadily increased into the 100–150-fc range. The extensive handbook issued by the Illuminating Engineering Society[65] considers lighting levels in practically every structure utilized by man (including dance halls: 5 fc), but is rather reticent about laboratories, mentioning these only very briefly in connection with schools and hospitals. The handbook recommends 100 fc (110 decalux in SI units) on the task for school laboratories; and 50 fc general illumination, 100 in close work areas for hospital facilities. Recommendations for offices cover a wide range, depending on the nature of the work, but 70–100 fc would seem to cover most activities. The radioisotope operation certainly justifies plenty of light since many of the techniques are those of microchemistry, particularly in the extensive use of micropipets.

Meaningful energy savings are possible if only the immediate work area is brightly lighted, with general illumination being at a much lower level, taking advantage of natural sunlight for the latter as much as possible. Light colors for any painted laboratory surface can be a help. Photoelectric devices for maintaining a fixed general light background are available, analogous to a thermostat in control of heat. Use of such approaches in the laboratory should help in reducing the annual electric bill, a not inconsequential operating cost item.

REFERENCES

1 W. C. McCluggage, "The AEC Accident Record and Recent Changes in the AEC Manual, Chapter 0529." USAEC Report BNWL-SA-3906 (1971); Anon, "Operational Accidents and Radiation Exposure Experience." USERDA Report WASH 1192 (1975).

2 D. C. Stewart, "Techniques of Handling Highly Active Beta- and Gamma-Emitting Material." In H. B. Jonassen and A. Weissberger, Eds., *Technique of Inorganic Chemistry* Volume III, Wiley-Interscience, New York (1963), pp. 167–258.

3 J. S. Swartout, Ed., "Radiochemistry Laboratories, a Symposium." Ind. Eng. Chem 41, 227–250 (1949).

4 Anon, "Laboratory Design for Handling Radioactive Material." Building Research Advisory Board Report NP 3875 (1952); NRC-NAS Research Report No. 3.

5   H. S. Coleman, Ed., *Laboratory Design*; *National Research Council Report on Design, Constru-tion and Equipment of Laboratories*. Reinhold, New York (1951).

6   H. F. Lewis, Ed., *Laboratory Planning for Chemistry and Chemical Engineering*. Reinhold, New York 1962: (a) D. R. Ward, "Low-level Radioisotope Laboratories," pp.156-170: (b) M. M. Solomon, "Facilities for Scientific and Technical Functions," pp. 103-144; (c) R. C. deWahl, "Materials of Construction," pp. 20-28.

7   J. J. Fitzgerald, *Applied Radiation Protection and Control*, Gordon and Breach, New York (1969). In two volumes. "Design of Nuclear Facilities" appears on pp. 354-410 of Vol. 1.

8   Anon, "Design and Construction of Laboratory Buildings": (a) "I. General Considerations." *Anal. Chem.*, 34 (10), 25A *et seq* .(1962); (b) "II. Academic Buildings." *ibid. 34*, (11) 23A *et seq* (1962); (c) M. H. Fairhurst, "Design of Laboratories Abroad." *ibid.*, *34* (13) 23A *et seq.* (1962).

9   M. G. Mellon, (a) "Some Trends in Planning Chemical Laboratories." *J. Chem. Educ.*, 52, 345-348 (1975); (b) "II. Structural Items." *ibid.*, 53, 114-116 (1976); (c) "III. Non-Structural Items." *ibid.*, 53, 454-456 (1976); (d) "IV. Composite Arrangement of Laboratories." *ibid.*, 54 195-198 (1977); (e) "V. Miscellaneous Trends in Building Materials." *ibid.*, 55, 194-197 (1978).

10 (a) Theodore Ruys, "37 Keys to Laboratory Design." Res./Develop. (December 1969) pp. 18-25; (b) J. W. Beyvl, "Untangling the Complexities of Laboratory Design." Am. Lab. (September 1969) pp. 37-41; (c) C. H. Wang, R. A. Adams and W. K. Bear, "Coordinated Design of Radioisotope Laboratories." Atomlight (New England Nuclear Corp.) No. 30 (July 1963); (d) D. Allison, "Places for Research." *Int. Sci. Tech.*, (September 1962), p. 20 *et seq.*; (e) N. S. Radin, "Design Principles in Building a Biochemistry Laboratory." Am. Lab. (January 1974) pp. 39-51.

11 IAEA (Chap. 1, Ref. 2), "Safe Handling of Radionuclides, 1973 Edition." IAEA Safety Series No. 1 (1973).

12 IAEA, "Radiation Protection Procedures." IAEA Safety Series No. 38 (1973).

13 American Institute of Chemical Engineers, "American Standard: Design Guide for a Radio-isotope Laboratory (Type B)." ANSI N 5.2 1963. American National Standards Institute (ANSI). ANSI, 1430 Broadway, New York, NY 10018.

14 International Organization for Standardization, "Fundamental Principles for Protection in the Design and Construction of Installations for Work on Unsealed Radioactive Materials." ISO/R 1710-1970(E), ANSI, New York (1970).

15 ERDA (Chap. 2, Ref. 44), "General Design Criteria (Handbook)." ERDA Manual Appendix 6301 (various dates).

16 Martin Sherwood, "Laboratory." *Chem. Tech.*, 2, 647 *et seq.* (1972).

17 Laboratory Investigation Unit (LIU), Department of Education and Science, Elizabeth House, York Road, London SE 1 79H, England. (A. J. Branton, Principal Architect): (a) A. J. Branton and F. P. F. J. Drake, "Available Furniture and Services for Education and Science." LIU Paper No. 6 (August 1972),: (b) M. V. Sinclair, K. Livingston and P. F. Bottle, "The Conversions of Buildings for Science and Technology: Part 2." LIU Paper No. 8 (September 1977); (c) F. Drake, "New Trends in Design for Science Laboratories." Lab. Equip. Digest (London) (February 1978).

18 R. G. Stein, *Architecture and Energy*, Anchor Doubleday, Garden City, NY (1977).

19 N. B. Garden, "Laboratory Handling of Radioactive Material." UN (Chap. 2, Ref. 5), First Geneva Conference, A/CONF 8/722, 7, 62-66 (1956).

20 IAEA (Chap. 1, Ref. 2). "Manual on Safety Aspects of the Design and Equipment of Hot Laboratories." IAEA Safety Series No. 30 (1969); Laboratory classifications are also given in IAEA SS No. 2 (Chap. 2, Ref. 38).

21 IAEA "Design of and Equipment for Hot Laboratories." Proceedings of the Symposium, Otaniemi, Finland, August 2-6, 1976. IAEA Proc. Ser. IAEA-SM 209/, STI/PUB/436 (1976); (a) C. M. Unruh, "Radiological Design of Hot Laboratories." pp. 183-190; (b) A. Beňadik, M Ďurćík and K. Martínek, "Safety Features in the Design of the Plutonium Laboratory of the Nuclear Research Institute." pp. 41-49; (c) J. D. Nixon and E. J. Chapman, "Ventilation and Filtration of Active Buildings." pp. 321-337; (d) C. Cesarano, N. Evangelisti, A. Natichionni, M. Lauro, G. Pugnetti and G. Vescia, "Design and Safety Criteria of the Alpha-Gamma Hot Laboratory of CNEN, Italy." pp. 209-222.

22 F. Rohloff, "Shielding of Fixed Storage Installations." *Eng. Compendium Rad. Shielding* (Chap. 4, Ref. 4). Vol. III, pp. 31-45.

23 L. G. Stang, Jr., "Hot Laboratory Equipment, Second Edition." USAEC Tech. Info. Service, Washington, D. C. (1958).

24 IAEA (Chap. 1, Ref. 2), "Symposium on the Safeguarding of Nuclear Materials." Vienna, October 20-24, 1975. IAEA Proc. Series IAEA-SM 201/, STI/PUB/408 (1976).

25 National Fire Protection Association (NFPA): (a) "Safety to Life from Fire in Buildings and Structures." NFPA Code 101 (1976); (b) "Fire Protection for Laboratories Using Chemicals." NFPA Code 45 (1975); (c) G. F. McKinnon and K. Tower, Eds., Fire Protection Handbook, 14th Ed. (1976). NFPA, 470 Atlantic Ave., Boston, Mass. 02210.

26 H. F. Lewis, "General Problems of Laboratory Design. *J. Chem. Educ.*, 24, 320-323 (1947).

27 D. C. Stewart, "A Large-Scale Facility for Chemical Research with Intensely Radioactive Material." In C. E. Crouthamel, Ed., *Progress in Nuclear Energy*, Series IX, Volume 3. Mac-Millan, New York (1963), pp. 237-265.

28 R. Spence, "An Atomic Energy Radiochemistry Laboratory—Design and Experience." UN (Chap. 1, Ref. 5), First Geneva Conference, A/CONF 8/438, 7, 39-43 (1956).

29 NRC (Chap. 2, Ref. 14), "Seismic Design Classification, Rev. 2." NRC Regulatory Guide 1.29 (March 1976); "Tornado Design Classification, Rev. 1." NRC Regulatory Guide 1.117 (April 1978).

30 American Society for Heating, Refrigerating and Air Conditioning (ASHRAE), *Handbook and Product Directory*. ASHRAE, 345 E. 47th St., New York, NY 10017. 1978 Applications Volume.

31 W. M. First, Ed., *Proceedings of the Thirteenth AEC Air Cleaning Conference. San Francisco, California, August 12-15, 1974*. Natl. Tech. Info. Service (Chap. 1, Ref. 4)): (a) M. Gonzales, J. Elder, and H. Ettinger, "Performance of Multiple HEPA Filters Against Plutonium Aerosols." pp. 501-525; (b) P. J. Magno, C. B. Nelson, and W. H. Ellet, "A Consideration of the Significance of Carbon-14 Discharges from the Nuclear Power Industry." pp. 1047-1055.

32 James Halutsky, "Estimation of Stack Height Required to Limit Contamination of Building Air Intakes." American Industrial Hygiene Conference, April 29, 1964. (Quoted in ASHRAE Handbook 1978, Ref. 33 above.)

33 W. G. Moroz, unpublished data. (Quoted in ASHRAE Handbook 1978, Ref. 33 above.)

34 S. J. Vachta, Argonne National Laboratory, personal communication.

35 IAEA (Chap. 1, Ref. 2), "Laboratory Training Manual on the Use of Radionuclides in Animal Research." IAEA Tech. Report Series No. 60 (1966).

36 J. F. Munce, *Laboratory Planning*, Butterworth's, London (1962). (Quoted in Ref. 39 above.)

37 D. Hughes, "Laboratory Ventilation and Fume Disposal." *Chem. Brit.*, 8, 288-292 (1972).

38 D.Hughes, "Design of Radionuclide Laboratories." *Ibid.* 4 63-66 (1968).

39 K. L. Powlesland and G. T. Saunders, "A Constant Volume Radioactive Hood." USAEC Report LRL 113 (1954).

40 J. P. Hughes and A. G. Jastrab, "Fiberglass Reinforced Plastic Gloveboxes for Plutonium Analytical Research." (Chap. 4, Ref. 2i), pp. 78-96 (1960).

41 B. M. Abraham and N. Bohlin, "Design of a Vacuum Frame Hood." USAEC Report ANL 4419 (1950).

42 W. M. First, Ed., *Proceedings of the Twelfth AEC Air Cleaning Conference. Oak Ridge, Tenn., August 28-31, 1972.* CONF 720823 (1973), in two volumes. Natl. Tech Info. Service (Chap. 1, Ref. 4): (a) J. R. Gaskell and J. L. Murrow, "Fire Protection of HEPA Filters by Using Water Sprays." pp. 103-122; (b) M. W. First, "Performance of Absolute Filters at Temperatures from Ambient to 1000° F." pp. 677-702; (c) C. L. Cheever, C. H. Youngquist, P. R. Hirsch, J. C. Hoh, D. S. Janetka and H. R. Fish, "Effects of High Level Gamma Radiation Exposure of HEPA Filters." pp. 638-645; (d) H. A. Lee, "Fire Protection in Caves, Canyons and Hot Cells." pp. 123-142.

43 W. M. First, Ed., *Proceedings of the Fourteenth ERDA Air Cleaning Conference, Sun Valley, Idaho, August 2-4, 1976.* CONF 760822 (1977), In two volumes *ibid.*: (a) D. A. Orth, G. H. Sykes and G. A. Schurr, "The SRP Sand Filter: More than a Pile of Sand." pp. 542-556; (b) C. A. Gunn and D. M. Eaton, "HEPA Filter Performance Comparative Study." pp. 630-661, (c) W. S. Gregory, K. H. Duerre, P. R. Smith and R. W. Andrae, "Tornado Depressurization and Air Cleaning Systems." pp. 171-193; (d) J. R. Gaskill, N. J. Alvares, D. G. Beason and H. W. Ford, Jr., "Preliminary Results of HEPA-Filter Smoke Plugging Tests Using the LLL Full-Scale Fire Test Facility." pp. 134-170; (e) W. J. McDowell, F. G. Seeley and M. T. Ryan, "Penetration of HEPA Filters by Alpha Recoil Aerosols." pp. 662-676, (f) R. A. Brown, Coord., "Summaries of Available Technology on Gaseous Effluent Control of Krypton, Iodine, Tritium, ^{14}Carbon, Ruthenium, NO_x, HCI and Particulates." pp. 1129-1133; (g) J. Bonenstigel, S. H. Dyer, M. Laser, St. Mastera, E. Merz and P. Morschl, "Separation of Fission Product Noble Gases Krypton and Xenon from Dissolver Off-Gas in Processing HTGR Fuel." pp. 1002-1016; (h) M. J. Stephenson and R. S. Eby, "Development of the FASTER Process for Removing Krypton-85, Carbon-14 and Other Contaminants from the Off-Gas of Fuel Reprocessing Plants." pp. 1017-1033.

44 C. A. Burchsted, J. E. Kahn and A. B. Fuller, "Nuclear Air Cleaning Handbook." USERDA Report 76-21 (1976).

45 AEC, "Revised Minimal Specification for the High-Efficiency Particulate Air Filter." USAEC Health and Safety Info. Bull., Issue No, 306 (March 1971).

46 U. S. Department of Defense, "Military Specification: Filter, Particulate, High-Efficiency, Fire-Resistant." USDOD Spec. MIL-F -51068C (June 1970).

47 Anon, "Standard for HEPA Filters." Institute of Environmental Sciences, 940 E. Northwest Highway, Mt. Prospect, Ill.

48 Anon, "Safety Standards for High-Efficiency Air Filter Units." Stnd. UL-586 (1976); "Safety Standards for Air Filter Units." Stnd. UL-900 (1977). Underwriter's Laboratories, 333 E. Tsingstin Road, Northbrook, Ill. 60062.

49 ASHRAE (Ref. 33 above), "Method of Testing Air Cleaning Devices Used in General Ventilation for Removing Particulate Matter." ASHRAE Stnd. 52-68 (1968).

50 ERDA, "Filter Unit Inspection and Testing Service Fiscal Year 1978." USERDA Environ., Safety and Health Info. Bull., Issue 342 (September 1977).

51 P. A. F. White and S. E. Smith, Eds., *High Efficiency Air Filtration.* Butterworth's, London (1964).

52 ACS (Chap. 1, Ref. 3), "Cleaning Our Environment, A Chemical Perspective," 2nd Ed. (1978); (a) p. 389; (b) p. 413; (c) p. 144.

53 L. M. Rubin and K. L. Miller, "A Solution to the Radioiodine Volatilization Problem." *Health Phys.,* 32, 307-309 (1977).

54 A. A. Moghissi and A. W. Carter, Eds., *Tritium,* Messenger Graphics, Phoenix, Ariz. (1973).

55 E. A. Evans, *Tritium and its Compounds,* Van Nostrand, Princeton, N. J. (1966).

56 V. R. Raaen, G. A. Ropp and H. P. Raaen, *Carbon-14,* McGraw-Hill, New York (1968).

57 J. R. Catch, *Carbon-14 Compounds*, Butterworth's, London (1961).

58 A. C. Stern, Ed., *Air Pollution, Engineering Control of Air Pollution*, Volume IV. Academic Press, New York (1977).

59 C. J. Barton, "Glove Box Techniques." In E. S. Perry and A. Weissberger, Eds., *Techniques of Chemistry, Laboratory Engineering and Manipulation*, Third Edition. Wiley-Interscience, New York (1978), Chap. V.

60 C. J. Barton, Glove Box Techniques." In H. B. Jonassen and A. Weissberger, Eds., *Techique of Inorganic Chemistry*, Volume III, Wiley-Interscience, New York (1963). pp. 259-333.

61 D. Hughes and R. Cullingworth, "Laboratory Fittings and Waste Systems." *Chem, Brit.*, 8, 470-474 (1972).

62 Anon, "Sweet's Catalogue File." Sweets Div, McGraw-Hill Info. Systems Co, 121 Ave of the Americas, New York 10029.

63 Kewaunee Scientific Equipment Corp., Adrian, Mich. 49221.

64 Speakman Safety Equipment, 301 E. 30th St., Wilmington, Del. 19899.

65 J. E. Kaufman, Ed., "IES Lighting Handbook." Illuminating Engineering Society, 345 E. 47th St., New York 10017. 5fh Ed. (1972).

$$\frac{\text{activity concentration per unit volume before treatment}}{\text{activity concentration per unit volume after treatment}} = \text{D.F.}$$

chapter 4

Shielded Facilities

Facilities where an operator on one side of a shield performs manipulations by remote means on radioactive material on the other side have been termed "hot cells," "caves," "hot laboratories" and "canyons." Such shielded enclosures are the subject of this chapter and may consist only of a small cubicle or composed of a series of huge interconnected rooms spread at different levels throughout a building. The main cell in one fuel reprocessing plant for example is 92 ft long by 22 ft deep by 43 ft high[1] and is itself dwarfed by some of the "canyons" at the Hanford and Savannah River governmental reservations.

The hot cell, in essence, is a room with extraordinarily thick walls. The operator outside those walls must either be able to see into the enclosure or have instruments to tell him what is going on inside; he must be able to manipulate materials in the interior with a delicacy and on a scale appropriate to his needs; he must be able to get both radioactive and nonradioactive objects in and out safely; he must be able to carry out repairs within the room in an emergency; in some cases he may have to have complete control of the composition of the atmosphere within the enclosure; he should have a very clear idea as to how cleanup and dismantlement will be accomplished; and, above all, he must be certain that the facility design and shielding are adequate for carrying out all these activities while at the same time protecting himself and everyone else from harm.

The best continuing and almost the only references to the state of the art of design and operation of shielded facilities are in the published proceedings of the "Hot Lab Group,"[2] whose meetings have now become international in character. This organization was formed in the early 1950s when a small number of American specialists felt the need for more than informal communication. The group survived a hand-to-mouth existence (in terms of having their proceedings published) until 1961 when they affiliated with the American Nuclear Society where they are now known as the Remote Systems Technology Division (RSTD). Much of their published material is heavily on the engineering side, but that is largely the name of the game.

Other extensive reviews of shielded-facility design and operation include those by Ferguson, Doe and Goertz[3a] and by Ring and Smith.[4a] Both cover the fundamentals in thorough fashion. The shorter paper by Unruh[5] has been previously cited. Proceedings of the international meetings sponsored by the European Nuclear Energy Agency[6] and by IAEA[7,8] contain directly relevant information.

4.1 GENERAL FEATURES

The zoning principle should be part of the design of even a small single shielded cell. The operator will be at a viewing window in front of the unit while carrying out manipulations. This "operating" area at worst should only be mildly suspect, that is, a Zone 2, with radiation exposure and possible contamination highly unlikely. The cell interior itself is Zone 4. Zone 3 must then be an enclosed area adjacent to and almost part of the cell where movements of materials in and out of the interior are carried out. Such access zones are variously known as "transfer areas," "maintenance areas," "service areas," "charging areas," etc.; but however named, they are the regions that serve as a buffer when the cell interior must be breached for transfers. This Zone-3 space of course encompasses the main access door to the cell. Contamination is an unwelcome visitor in a transfer area, but there is not much surprise if it occasionally occurs. Appropriate advance precautions and subsequent cleanup measures must be taken.

The ventilation pattern is obvious. The air must move from the clean operation side to the highly suspect transfer area, into the cell and out through the various air-treatment devices. Some specialized aspects of hot-cell ventilation will be considered later.

Multicell-facility layouts are essentially all variants of two basic designs, "in-line" or "back-to-back" as shown in Figure 4.1.[9] The fundamental pattern is however sometimes hard to discern since the transfer area may be on top of or below the cell bank or itself shielded and at first glance appearing to be part of the high-level working area. The cells may be on several different levels and connected by hatches or tunnels, but this is basically an in-line configuration.

Other variations include the L-shaped hot laboratory bank at Fontenay-aux-Roses in France[10] and the T of the TURF (Thorium-Uranium Recycle Facility) at Oak Ridge.[11] The Russians [7a] have proposed a circular facility but it is not clear that it was ever built. The HFEF/S (Hot Fuel Examination Facility, South) at Argonne-West in Idaho may be truly unique, being in the shape of a hand mirror with a rectangular air cell as the handle and a sixteen-sided controlled-atmosphere cell as the mirror.[12] Its sister facility, HFEF/N,[13] superficially appears to be much simpler, being in the shape of a rectangle with twenty-one working stations around the periphery, but is much newer and possibly the most advanced facility of its kind in existence. The six sets of back-

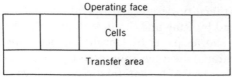

In-line arrangement

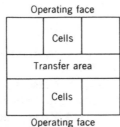

Operating face

Back-to-back arrangement

Figure 4.1 Basic layouts for multiple-cell facilities.

to-back cells at the Windscale Plant in England are set at right angles to a common transport corridor across the back, giving the complex the appearance of a comb.[14]

Both the basic in-line and back-to-back plans are conservative of space, reduce the amount of shielding needed because of the shared walls between cells, allow ready establishment of zoning patterns and simplify the problem of transferring materials from cell to cell, particularly for the in-line design where items can be passed from one end to the other through interior-wall penetrations. There are of course also disadvantages. Manipulation and viewing into an individual cell can only be from one side except for the end units, the types of access doors that can be used are limited and the nearness of each cell to its neighbor can create problems in finding room for out-of-cell instrumentation. The British[4a] have devised an interesting variation of the in-line concept. Each cell in a line is free standing with the units interconnected by transfer tunnels and pneumatic "rabbits" for materials transfer. Viewing can then be had from several sides into each cell and space is available for any external equipment.

Figure 4.2 shows a variation[15] of the back-to-back plan in the form of an H, a design that overcomes many of the shortcomings of the basic pattern. Viewing and manipulation are possible from three sides of every cell and each can be used independently of the others. The interior corridor acts as a radiation lock analogous to the air locks discussed in Chapter 3 since a cell may be opened with protection for the external areas still being provided by the main doors at the facility entrance.

The analytical chemistry cells at Oak Ridge[16] and the Savannah River High Level Caves[17] are good examples of the basic in-line design, while the High Activity Handling Building at Harwell[18] and the alpha-gamma cells at Los Alamos[19] are good representatives of the basic back-to-back pattern. Many

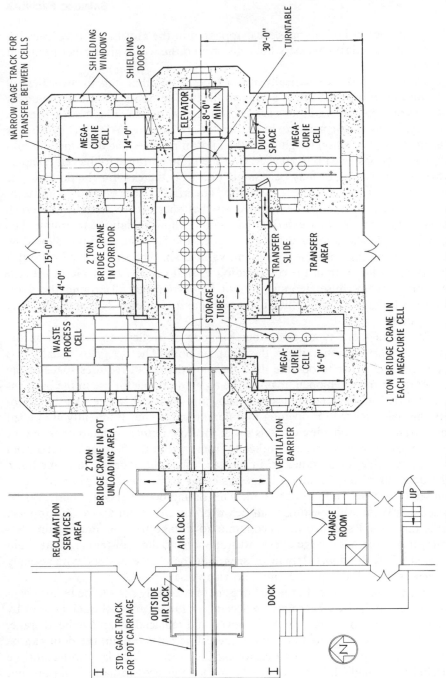

Figure 4.2 Hot-cell facility in the form of an H. (ANL Photo 120-4190 Rev.)

other examples of both designs have appeared in the Hot Lab Proceedings.[2] A survey[21] of analytical chemistry hot cells around the world gives other pertinent citations.

Cell access doors are basically of two types, those that slide to cover the cell opening and those that act as a plug to fill it. The doors are usually made of steel or concrete, but variations are seen such as concrete or lead brick encased in a steel lining. Some units are huge, the "doors" for the LAMPF (Los Alamos Meson Physics Facility) weighing in the kiloton range.[22] The door may be split into two sections as with the one serving the main entrance to the cell complex shown in Figure 4.2, or may be a single unit as is the case for the doors to the individual cells shown in the same figure. The mating sides of the two halves of the split-door types must be stepped so that there is no straight-line crack between the interior and exterior that would allow radiation streaming. The same principle also applies to the plug types, the face and its frame should be stepped in the fashion of a large bank-vault door. The sliding type of unit is always made larger than the cell opening so that there is an overlap to reduce streaming, but because stepping cannot be carried out all around the door, radiation leakage is more difficult to control and supplementary shielding may be required.

The sliding-door types can either be hung on trolleys on I beams on the ceiling or operated on rails in the floor or both. The design is usually less expensive and easier to operate than the plug doors but requires space at the side of the frame when open, making the type unsuitable for use in the basic in-line and back-to-back configurations shown in Figure 4.1. The plug types obtain the space needed for opening in different ways. They frequently are pulled back from the cell opening on rails set in the floor, either by hand or by motor drive. They may be raised hydraulically from a pit in the floor or lowered from the top by cables or a crane, or, in the case of a split door, by both methods. The door can also be hung on hinges as either a one piece or split unit, again similar to a bank-vault installation. The sliding-door version could also be raised across the cell opening from below or lowered from above, but in most cases this would sacrifice any economic advantage without much gain elsewhere unless there is a large overhead crane serving the transfer area that could be used to lift the doors. It will be seen that the choice of door type strongly influences the cell-complex design and vice versa.

Large-opening access to the cell does not necessarily need to be by means of a door but can be by a hatch giving access to a tunnel under the cell floor or by a stepped ceiling plug that can be removed and replaced by an overhead crane.

If the level of expected activity in the cell is not too high, the door can be eliminated completely and a labyrinth used for access. Such labyrinths must be carefully designed[23] because of the hazard of reflected radiation and are not suitable if the radiation source within the cell is such as to produce loose contamination. Labyrinths are most useful when the radiation source is intermittent and the cell must be frequently entered. Examples might be the housing of an x-ray machine or a modest-level ^{60}Co facility where the encapsulated sources could be remotely placed in shielded storage between irradiations.

The cell side of the access door is susceptible to contamination which may cause problems in the transfer area when the cell is opened. Solution of this problem is frequently by use of a second inner door that may be nothing more than a sheet of plastic or an aluminum unit similar to the ordinary household screen door. Contamination leaking out into the transfer area around the sides of the door is a consideration, particularly for the sliding types. Seals are accordingly used on both the outer-shielding and inner-containment barrier doors. These seals can be of different levels of complexity, one ingenious type being a long flexible tube around the door frame that can be inflated with compressed air when the door is in the closed position. Such inflatable seals find other uses in transfer locks, etc.

Cell designs are often differentiated as either "beta-gamma" or "alpha-gamma."[3a] The dividing line is sometimes rather blurry, but in the former type the work to be done in the cell involves very low or zero quantities of the more biologically damaging and generally much longer-lived alpha emitters. The contrast between the two cell designations is thus comparable to that between a radioisotope hood and a glovebox, and in fact the shielded box shown in Figure 3.3 of the last chapter could be considered to be an alpha-gamma facility. In a beta-gamma cell, as in a hood, the chief control of contamination spread is the ventilation system. The cell is kept at negative pressure so all air flow is inwards and access holes, shielding doors, manipulator penetrations, etc., are not necessarily completely sealed. In contrast, the alpha-gamma design is buttoned up tightly with all openings to the outside sealed, although the ventilation pattern of course stays the same. Many of the alpha-gamma facilities also have controlled atmospheres, partly for fire control (some of the heavy metals such as plutonium are pyrophoric in powdered form) and partly to avoid surface oxidation of materials. Nitrogen, helium or argon are the gases usually employed. Nitrogen is the least costly, but can undergo chemical reaction with some surfaces. Argon and helium are inert chemically but the former, the most expensive, is preferred because its high density means that any traces of lighter gases present will tend to float to the top and away from the working surface. These gases are used in closed systems, that is, they are circulated into the cell, withdrawn, cleaned by filtration, etc., and reused with only a limited amount of makeup gas added when necessary. A controlled-atmosphere design naturally places additional emphasis on air-tight seals and of course very considerably complicates the engineering needed to get materials in and out without diluting the cell gas. Transfers are usually made by means of a lock which in its simplest form could be a pipe through the wall with an air-tight door at either end. The object being transferred is placed in the pipe, the outer door closed, the lock purged with cell gas, and the inner door then opened so that the object can be moved into the cell. Since the questions of shielding violation and contamination spread must also be considered, transfer locks are very rarely as simple as described.[24]

A beta-gamma cell can be used with high levels of alpha activity by employing "containment boxes"—essentially gloveboxes within the cell with manipulators in booting substituting for human arms in gloves. Such containment boxes can be of a size to allow several to be used simultaneously in a larger cell or a single unit may effectively occupy the entire cell volume.

The boxes may be constructed of various materials such as stainless steel, aluminum, plastic sheet or even plywood; the material chosen usually depending on the projected working life of the unit. If this time is of the order of weeks or months, light and easily demolished materials should be chosen in order to simplify the dismantlement process at the end of the experiment when the box and its remaining contents must be disposed of as active waste. Figure 4.3 shows a containment box built for transplutonium element research. The photo was taken from inside the cell before work began and shows the back side of the box. The transparent side, front and back panels are of plastic sheet ($\frac{1}{4}$-in. Lucite) sealed into commercially available aluminum or vinyl plastic extruded channels. The floor is a thin aluminum sheet sealed into the channels and supported by plywood backing. The slave arms of the Model 8 manipulators coming through the wall at the top of the figure are contamination protected in booting formed in flexible vinyl sheeting that is in turn sealed to the box to form an integral roof. The gloves seen on the back and side of the box are

Figure 4.3   An in-cell containment box. (ANL Photo 121-1504.)

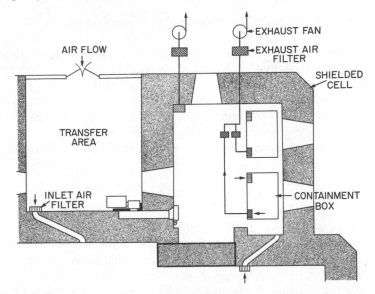

Figure 4.4   Air-flow pattern in a high-activity level cell. (ANL Photo 120-6196.)

primarily for the later cleanup and disposal operation. The manipulators will be used at the end of the experiment to eliminate as much active material from the box as possible and to decontaminate it to the point where operators using the gloves can finish the job. A plastic bag for waste is seen hanging from the transfer port at the right. The operating-side viewing window is visible throught the box and there is a second window on the opposite side of the cell in back of the photographer (Figure 4.4). An operator at this second window can transfer materials in and out of the lock with an extended-reach manipulator.[25]

The shielded cell is itself at lower pressure than the operating area and the box interior negative to the cell atmosphere. Air is taken into the box through the small HEPA filter at the upper right of the box and is exhausted through the double-filter unit at the left. The exhaust air is carried by flexible hose to a second filter permanently installed within the cell and then finally discharged through a third filter in the fan loft. The exhaust thus passes through three HEPA filters in series. The overall air pattern is shown in Figure 4.4.[26]

Containment boxes for long-period use are almost always of metal construction and elaborately engineered.

4.2   SHIELDING: ALPHAS AND BETAS

As discussed in Chapter 2, charged particles are slowed down and stopped relatively rapidly because of electromagnetic interaction with the atomic fields of the target (shielding in this case) so have comparatively little penetrating power. The uncharged radiations such as x-rays, gammas and neutrons on the other hand must have essentially direct encounters with specific atoms before their passage through material is affected. Accordingly, these are the "penetrating" radiations.

Some of the extremely heavy transplutonium isotopes and a few short-lived members of natural-decay chains emit alpha particles of greater than 6.5 MeV energy, but an encounter with these is rare in usual radioisotope work. Assuming a 5-MeV value as reasonably typical, the penetration of the particle in tissue would be about 45 μ,[27] and of course less in denser materials. For comparison, the protective layer of the human skin is about 70 μ.

The penetration value cited above was calculated from the Bragg-Kleeman equation:

$$R_s \ = \ \frac{3.2 \times 10^4 \, R \, A^{1/2}}{\rho}$$

where R_s is the penetration depth in centimeters of an alpha in a solid material and R is the range of the same alpha in air. A is the atomic number of the material and ρ is its density. The equation is somewhat approximate and its use of course requires that the range of the alpha of interest in air be known, information that is given in Figure 4.5. [The ranges in the figure are presented

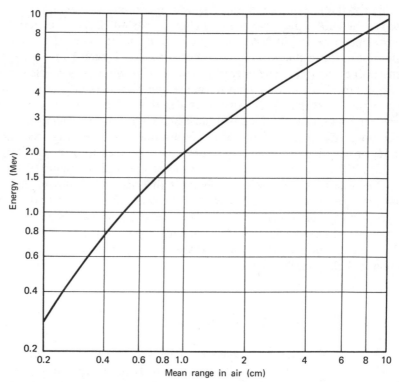

Figure 4.5   Range of alpha particles in air. (Ralph E. Lapp and Howard L. Andrews, *Nuclear Radiation Physics*, 3rd ed., ©1963, p. 117. Reprinted by permission of Prentice-Hall, Inc., Englewood Cliffs, N.J.)

in terms of centimeters. The unit g/cm² is also frequently used in graphing charged particle absorption data and is obtained by multiplying the absorber thickness in centimeters by the material's density expressed as g/cm³.]

Shielding of isotope alphas produced externally to the body is thus not a great problem. Beta-ray shielding can be somewhat more complicated in some situations.

Considerable beta-ray scattering occurs as the particles pass through matter, and, as pointed out in Chapter 2, the apparent range of the beta will be generally quite different from its total path length. Since the betas are also emitted with a continuous energy spectrum, theoretical analysis of the absorption process becomes extremely complicated. Luckily the combination of the scattering and the continuous-energy-spectrum phenomena fortuitously produces an apparently exponential absorption for betas of a given energy over a considerable range of absorber thicknesses. This is expressed as

$$I = I_{0}e^{-\mu x},$$

where I_0 equals the original intensity of the beam and I is the intensity at depth x in the absorber; μ is defined as the absorption coefficient for that particular material and beta energy and is expressed as cm⁻¹ if x is in centimeters. The ratio μ/ρ (or the mass absorption coefficient where ρ is the density of the absorber) is nearly independent of the nature of the absorbing material.

The range of a beta particle is primarily dependent on the mass of the material it traverses, that is, the atomic composition of the material has relatively little effect. The range is thus roughly inversely proportional to the density of the absorber. A number of somewhat more precise empirical formulations has been published. Glendenin[28] gives

$$R = 0.542E - 0.133 \qquad (E > 0.8 \text{ MeV})$$

and

$$R = 0.407E^{1.38} \qquad (0.15 < E < 0.8 \text{ MeV}).$$

In this case R is in g/cm² and E is the maximum energy of the beta in MeV. Figure 4.6 presents shielding data for betas in another format.

Beta energies greater than 2 MeV are rather uncommon among most of the isotopes usually encountered. It will be seen from Figure 4.6 that even at that energy level the particles will be stopped by $\frac{1}{10}$ in. or less of most shielding materials. There is however a complication that can be important in some cases. Electromagnetic theory requires that a moving electric charge radiate energy when it is decelerated. A beta particle as it slows down, within the source itself due to self-absorption or in external shielding, gives off this "braking" radiation or "bremsstrahlung" which appears as a continuous x-ray spectrum. The betas themselves also have a continuous energy spectrum as they are emitted from an isotope, with the mean energy usually about one-third that of the maximum value as given on the isotope charts. The bremsstrahlung spectrum will be in the same energy range (but not as a matching curve) but at

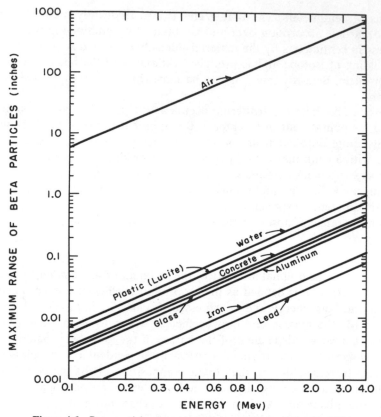

Figure 4.6　Beta-particle ranges in various materials. (From Ref. 29.)

much lower intensity as can be seen from the examples given in Table 4.1. The intensity of the bremsstrahlung increases approximately with the energy of the beta particle and with the square of the atomic number of the absorbing material. The phenomenon is thus primarily a problem when working with large quantities of a high-energy beta emitter and heavy-element shields such as lead. Bremsstrahlung shielding is obviously an important consideration around a beta accelerator since the x-rays produced can be in the energy range of hundreds of MeV for a high-powered machine.

If the isotope source is large or bulky the braking radiation will largely be produced within the material itself due to self-absorption. This can lead to interesting situations in isotopic heat source work as shown in Figure 4.7 adapted from Ref. 31. Promethium-147 for example emits only a weak ($0.23 -$ MeV) beta, but when unshielded the bremsstrahlung dose delivered by a Pm_2O_3 source is much more substantial than that from the direct radiation. The dose originating in a similar unshielded ^{90}Sr-^{90}Y source is essentially all from bremsstrahlung, the preponderance due to the unusually energetic (2.27-MeV) ^{90}Y beta.

Table 4.1 Bremsstrahlung from Beta Emitters

| Nuclide | Beta-Particle-Energy | | Bremsstrahlung |
	Maximum (MeV)	Average (MeV)	(MeV/beta)
^{106}Rh	3.54	1.515	1.29×10^{-1}
^{90}Y	2.27	0.944	2.81×10^{-2}
^{90}Sr	0.545	0.201	1.41×10^{-3}
^{147}Pm	0.23	0.067	2.02×10^{-4}
^{171}Tm	0.097	0.029	3.13×10^{-5}

Source: Adapted from Appendix 11, Ref. 30.

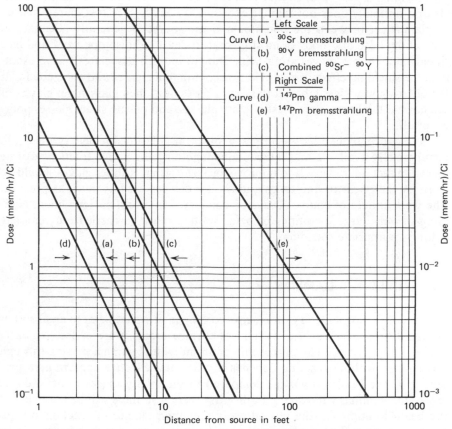

Figure 4.7 Bremsstrahlung contributions to delivered dose. (Adapted from Ref. 31.)

The Radiation Health Handbook[32] cites various rules of thumb. A few paraphrases with application to beta-ray emission are:

1 The range (R) of beta particles in gm/cm^2 is approximately equal to one-half the maximum energy (E) in MeV, that is, $R = E/2$.

2 (R) for betas in air is about 12 ft/MeV—a 3-MeV particle will have approximately a 36-ft range.

3 Betas passing through light materials such as water, glass or aluminum dissipate less than 1% of their energy as bremsstrahlung.

4 The dose rate in rads/hr at a distance of 1 ft from a point source of beta radiation (neglecting self- and air absorption) is approximately 300 times the number of curies in the source.

5 A beta particle must have an energy in excess of about 70 keV to penetrate the protective layer of the skin.

Shielding a pure beta emitter is thus best done with lighter-element materials and normally does not require excessive thicknesses. A transparent plastic or glass sheet between the operator and the source is generally adequate to protect the head and body in most laboratory-level radioisotope work and face shields and rubber gloves are a help. Dunlap and his associates[33] have examined the shielding effectiveness of various ophthalmic lenses and concluded that the ordering was glass laboratory safety glasses best, then glass street glasses of the normal thickness, then plastic safety glasses, and plastic street glasses poorest of all.

Avoidance of hand exposure is a problem. Plastic-shielded syringes are commercially available and the containers holding the active fractions can be placed in secondary vessels to improve local shielding. Rubber gloves should be worn by the operator and tongs and similar devices used as much as possible. The work should of course always be conducted in a radioisotope hood. Handling of very large quantities of beta emitters must be carried out remotely behind appropriate shielding.

4.3 SHIELDING: GAMMAS

A beam of monoenergetic gamma-rays coming in at right angles to an absorber or shield first "sees" a bead curtain of surface atoms more or less regularly spaced in opposition. The probability of an individual gamma passing through this curtain without reacting depends on the total effective capture area presented by the beads (atoms) as compared to the empty spaces in between. A few of the gammas will react with this first curtain so there will be fewer left in the beam to approach the second array, but their chance of reaction still remains the same. The pattern is continuously repeated—each new curtain of atoms will remove a certain fraction of those gammas that have survived to that point. The process can thus be expressed in the same exponential form as given in the last section for betas:

$$I = I_0 e^{-\mu x}$$

except that in the gamma case the pattern is not fortuitous.

This very simplistic bead-curtain analogy can help with another point. The ratio of total capture area to open space in a single curtain will depend on the number of beads in a given target area (a function of the absorber's density) and on the *apparent* target size of each individual atomic bead. This size is expressed as a cross section (σ), the unit for which in nuclear physics circles is the barn (10^{-24} cm^2).

Exponential disappearance is the fundamental pattern for gamma absorption in a shield but many qualifications have to be made, simply because atoms are not hard beads hung on invisible strings but highly complex interactive and reactive systems. An incoming gamma can undergo reaction in any of four locations within the atom: the atomic electron cloud, with the nucleons in the nucleus, in the electric field surrounding the nucleus and the electrons, or in the meson field surrounding the protons and neutrons in the nucleus. In each case the reaction in theory can then proceed in any of three directions: the gamma can be absorbed and completely disappear, it can undergo elastic (coherent) scattering or it can undergo inelastic (incoherent) scattering. A total of twelve overall reaction patterns are thus theoretically possible, most of which have been observed.[36] Each atom thus has twelve different apparent gamma reaction cross sections.

(If an atomic system recoils as a whole under the impact of a gamma its internal energy is not changed and the gamma scatters away elastically. If the impact causes an atomic particle to recoil with respect to others, the internal energy of the system is increased and the gamma is scattered inelastically with its own energy correspondingly reduced.)

Three of the twelve possible reaction mechanisms for the gamma predominate for most practical purposes and therefore are of primary interest in shielding. The photoelectric process is a true absorption. The gamma disappears, its energy going into the ejection of an electron from the target atom. The photoelectric effect is the predominant reaction mechanism for low-energy gamma rays, especially for high-Z (atomic number) materials. In Compton scattering the photon is inelasticallly scattered and loses energy which is transferred to an electron which recoils out of the atom. The Compton effect is predominant for 1–5-MeV gammas in high-Z materials and even more so over a much wider energy range in low-Z targets. The third important reaction mechanism is pair production. The photon disappears and its energy produces an electron-positron pair which originates out of the space around the atomic nucleus. The threshold for formation of an $e^- - e^+$ pair is 1.02 MeV so pair production cannot occur at all with gammas of less than that energy, but becomes predominant at high photon energies, particularly in high-Z materials.

The relative importance of each of these three major processes thus varies with both the energy of the photon and the nature of the element or elements in the target. Figure 4.8 shows the patterns in a light element (Al) absorber as compared to a heavy element (Pb) target. Because the photoelectric effect is of primary importance at one end of the energy range and pair production at the

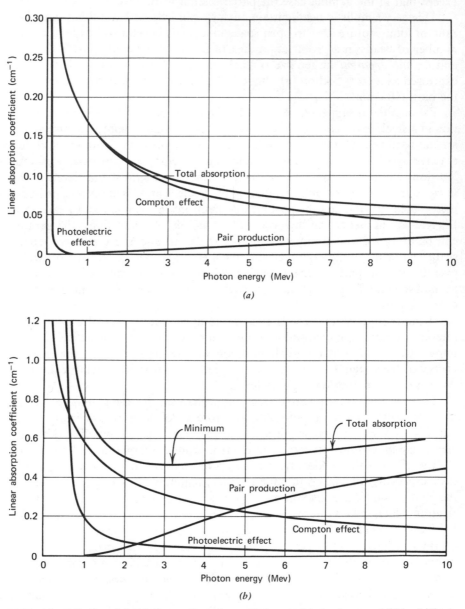

Figure 4.8 Relative values of photon absorption coefficients in (a) aluminum and (b) lead. (From Ref. 34, Courtesy American Physical Society.)

other, the total summed curve for each element has a minimum for the overall absorption coefficient at some point in between. This minimum comes at rather high energies for the lighter elements (about 22 MeV for Al).

Two full, closely packed pages describing the units and notation used in shielding calculations are given by Hubbell and Berger.[4b] Only those most usually seen will be discussed here.

The "narrow beam linear attenuation coefficient," μ, is also encountered in the literature as the "linear absorption coefficient" or as the "macroscopic absorption coefficient." The quantity is an expression of the probability that a gamma of a specific energy will react within a given volume of the shielding element or material in question. The coefficient is related to the cross section through:

$$\mu = n\,\sigma,$$

where n is the number of atoms in that volume. A second frequently seen term is the mass absorption coefficient, μ_m, or Σ, obtained as described in the last section by dividing μ by the absorber density. The mass absorption coefficient is particularly useful in dealing with heterogeneous materials such as concrete or with substances in different density states (water versus ice). The coefficient is the fraction of the beam removed by unit areal density (1 g/cm^2), regardless of the thickness needed to obtain that density, that is, the beam must encounter a prescribed number of atoms no matter how much depth this requires.

The mean free path (mfp) is the average distance traveled by a photon before absorption and is equal to $1/\mu$ and is equivalent to the relaxation length, λ, for monoenergetic photons. The basic exponential equation indicates that this is the distance in the absorber required to attentuate the beam by a factor of $1/e$ (0.367). The half-value layer (HVL) and the tenth-value layer (TVL) quantities are particularly useful for quick estimates in everyday work. Both are given in terms of length and indicate the thickness of shield needed to attenuate the beam by factors of 2 and 10, respectively.

All the quantities defined in the last paragraphs are interrelated through μ and the shield density. These relationships are summarized in Table 4.2. The mean free path, λ, HVL and TVL can also be expressed in terms of g/cm^2 if μ_m is substituted for μ in the appropriate equations. Metric units are indicated, but any other system could be used if care is taken to be certain that the units are internally consistent throughout.

Acquisition of shield absorption data has necessarily been by a combination of theoretical and experimental approaches and the numbers are still being modified as new information is obtained. Much of this effort has been undertaken by the National Bureau of Standards sponsored by the National Standard Reference Data System. The first modern systematic compilation covering the energy range from 10 keV to 100 MeV (but not for all of the elements) was done by White.[35] She later modified some of her data and republished under her married name of Grodstein.[36] Further refinements were made by McGinnies,[37] Davisson,[38] Hubbell,[39] and Hubbell and Berger.[4b] Storm and his associates

Table 4.2   Relationships between Shielding Terms

Quantity	Symbol	Unit	Relationships
Cross section	σ	cm^2	
Linear absorption coefficient	μ	cm[^{1}	$\mu = n\sigma$
Density	ρ	g/cm^3	
Mass absorption coefficient	μ_m	cm^2/g	$\mu_m = \mu/\rho$
Mean free path	mfp	cm	mfp $= 1/\mu$ $I/I_o = 0.367$
Relaxation length	λ	cm	$\lambda = 1/\mu$ $I/I_o = 0.367$
Half-value layer	HVL	cm	HVL $= 0.693/\mu$
Tenth-value layer	TVL	cm	TVL $= 2.303/\mu$ TVL $= 3.323 \times$ HVL

Source: Reference 4b.

at Los Alamos published [40] photon cross sections for all of the elements from hydrogen through fermium over the 1 keV–100 MeV range, a compilation that was later updated by Storm and Israel.[41] Walker and Grotenhuis[42] published constants for the various types of concrete used in shielding. A number of other reviews and tabulations are cited in Ref. 4b.

Data from these various sources have been extensively reproduced, although sometimes in graphical rather than tabular form or for a restricted number of elements and mixtures or for a limited photon energy range. Such secondary information sources include the *Reactor Handbook*,[3c] the *Radiological Health Handbook*,[32] several of the IAEA handbooks,[43,44] Courtney,[45] Schaeffer,[46] Steigelmann,[31] Rockwell,[47] Goldstein and Wilkins,[48] and Capo.[49] Figure 4.9 shows a graphical representation for some common shielding materials, taken from Steigelmann.

These "narrow-beam" attenuation coefficients must however be applied with caution in practical work because they refer to a carefully defined and rather artificial geometrical situation for which the simple exponential law of absorption holds true. In this model it is assumed that the photons pass through a collimating slit so that they encounter the shield at right angles and that they are again collimated before striking the detector on the other side of the shield. The instrument then effectively sees only the surviving "uncollided" gammas from the original beam. There are of course extensive scattering reactions as the beam passes through the absorber, but these are not measured by the detector and from its point of view have been absorbed.

Many of the scattered photons will be absorbed within the shield but others will come out at odd angles and be measureable if the collimated detector is replaced by one that integrates the total flux coming to it from all directions.

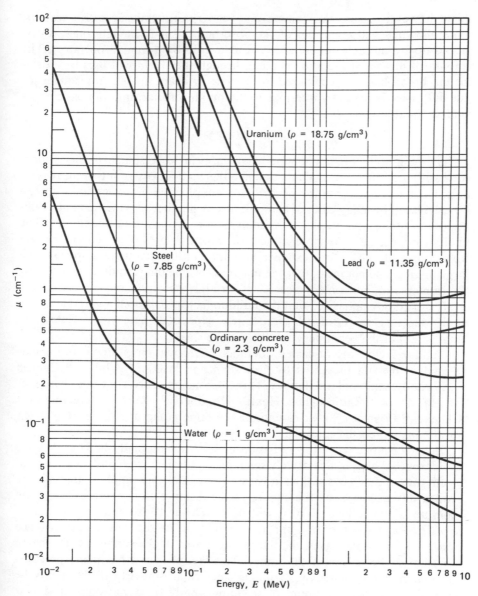

Figure 4.9 Gamma linear absorption coefficients versus energy. (From Ref. 31.)

The narrow-beam model is also generally unrealistic since the radiation from the source is not normally neatly collimated and thus first strikes the inner shield surface at many different angles. This is the much more typical "broad-beam" situation. The simple exponential equation will hold approximately if the shield is thin; the photons will on the average only be scattered out of the beam but once and at large enough angles to miss the detector. In a thicker shield however, where the gamma may undergo several successive scatters, some of which will come out of the absorber at small enough angles to be picked up by the detector (or received as a dose by a human operator). The amount of measured radiation passing through the shield will thus be greater for the broad-beam as compared to the narrow-beam model. Adjustment for this difference is made by inserting B, the buildup factor, into the basic exponential equation:

$$I = I_0 B\, e^{-\mu x}.$$

The buildup factor would superficially appear to be similar to the "rubber constant" used by some students to make their laboratory results seem reasonable, but the factors arise from largely known physical processes and thus can be calculated in many cases. There are also a number of experimental evaluations, so tables and graphs of buildup factors are again in abundance. The values can become quite large, almost 1000 for a 0.25-MeV gamma in water,[3c] for example. Because of the exponential nature of the attentuation process such large numbers do not mean that the thickness of shielding required increases by the same factor, but some increase will certainly be necessary.

It should be emphasized that the buildup factor, B, is itself a variable whose value depends on the depth (frequently expressed in terms of relaxation lengths, μx) of the beam penetration into the shield.

Buildup factors are discussed, methods of calculation outlined, and extensive tables of data presented by Goldstein and Wilkins,[48] Chilton,[3c] Capo,[50] Kuspa and Tsoulfanides,[51] Trubey,[52] etc.. Secondary sources include the Reactor Handbook,[3] Radiological Health Handbook,[32] Rees,[53] Steigelmann,[31] Schaeffer,[46] Courtney,[45] etc.; essentially the same group of references as given for attenuation-coefficient data.

In radioisotope work the radiation emitter can normally be considered a "point source," that is, a relatively small mass radiating in all directions (isotropically) on a 360° basis. The inverse-square law thus applies, that is, the gamma flux decreases as the square of the distance from the source. (If the exposure is 100 R/hr at 1 ft, it will be 1 R/hr at 10 ft.) The source can be considered as the center of a sphere of radiation. As the radius—the distance from the source—increases, the flux per unit area of the surface of the sphere decreases as the square of the radius. This fact of course has many practical implications for radioactivity handling since small sources can frequently be manipulated with a minimum of shielding simply by using tongs or similar devices. Bare (unshielded) nuclear reactors have been remotely operated in open fields in certain development programs and safety insured primarily by interdicting a large surrounding area to human entry.

Many sources are neither of the point nor the narrow-beam varieties but can be of all sorts of awkward shapes such as a reactor fuel rod, irradiated hardware, or the reactor core itself. Shielding calculations can be approximated for many of these geometries by considering that the radiation is concentrated as a series of point sources over the surface. Rockwell[47] considers line, disk, infinite slab, truncated right circular-cone, and cylindrical and spherically shaped sources in considerable detail. The effect of the shape of the source is also treated by Glasstone and Sesonski[54] and by Moteff[55]; and applicable formulas are given in the *Reactor Handbook*,[3] by Courtney[45] and by Schaeffer.[46] Physically very large radiation sources are fortunately usually problems for the reactor engineer rather than the scientist; as is a second area with geometrical overtones, that of radiation streaming through ducts and other penetrations embedded in large shielding walls. Treatment of this latter question is given in the various reactor-oriented references previously cited. Many papers on radiation streaming through openings of various shapes have appeared in the Transactions of the American Nuclear Society.

The above has been a very brief review of the background for analytical studies of gamma shielding. Mention of the Radiation Shielding Center at Oak Ridge[56] should be made at this point. This group was established in 1962 and "acquires, selects, stores, retrieves, evaluates, analyzes, synthethizes and disseminates information on shielding and ionized radiation transport. Computer codes and computer-readable nuclear data are treated as an integral part of this work." The comparable European Shielding Information Service, located in Ispra, Italy, was established in 1972.[57]

4.4 SHIELDING: NEUTRONS

Shielding against neutrons would appear to be primarily a problem for the reactor engineers rather than for the bench scientist, but this is not entirely true. Those individuals working with substantial quantities of alpha emitters have to be aware of (α, n)* generated neutrons and of the heavy-element isotopes having short spontaneous-fission half-lives, ^{252}Cf being the best-known example.

Free neutrons can only originate from atomic nuclei, and are usually ejected at energies in the 1–10-MeV range. This means that they are initially moving at as much as one-tenth of the speed of light and for obvious reasons are termed fast neutrons. Having no charge, neutrons are highly penetrating into matter, being stopped or slowed only by direct encounters with individual atomic nuclei in the target. Even in a complete vacuum the neutron could not however go on forever since it itself is a radioactive species decaying with a 10.6-min half-life to form a proton-electron pair. This fact is of little practical importance in the normal situation since the neutrons are destroyed in a very much shorter time by absorption in matter.

Bringing about this absorption as quickly as possible is the goal of the shielding engineer, but its accomplishment is not simple: ". . .calculation of neutron attenuation is exceptionally difficult."[3d]

Neutrons basically react with matter in one of two ways—they are either scattered by or absorbed into any nucleus with which they collide. The probability that a fast neutron will disappear (expressed as an absorption cross section) upon its first encounter with a nucleus is almost nonexistent, so the particle must be made to lose its energy (slowed) as rapidly as possible to make its capture more probable, which in practical terms comes down to selection of the optimum shielding composition. This slowing process comes about through scattering reactions, a lengthy series of which may be required before the neutron energy is reduced to the thermal or "slow" range (0.025 eV at 22°C) where absorption cross sections generally increase dramatically. A thermal neutron will be moving at speeds comparable to those of the atoms about it. If the target is artificially cooled to near absolute zero, "cold" neutrons are eventually produced.

The scattering reactions undergone by neutrons can again be either elastic or nonelastic. In the latter case the neutron is actually very briefly absorbed in an atom to form a highly excited "compound" nucleus. There are various alternatives for this compound nucleus to adjust for the unwelcome influx of energy, the resulting reaction ruled by a set of probabilities depending in a very complex way on the energy-level states of the target atom and the exact neutron energy. Fission may occur in certain heavy-element-target situations, or photons, charged particles, or neutrons may be emitted to accomodate. This last is basically the inelastic-scattering mechanism of interest here—a neutron comes in, a neutron of degraded energy comes out. This phenomenon can occur only if the incident neutron carries enough energy to raise the target nucleus to its first-excited-state level (generally above 0.1 MeV). The energy-level spacings in heavier atoms are characteristically such as to make this more probable. In other words, neutrons above 0.1-MeV energy are most efficiently slowed in heavy-element shielding by inelastic scattering, but this advantage is very rapidly lost once the particle's energy is below a threshold value which can be quite high in the lighter elements. The slowing process from then on in is by elastic scattering.

While elastic scattering may involve formation of a compound nucleus and subsequent neutron emission, the phenomenon can be regarded as a billiard ball type of collision and the laws of classical physics apply. A billiard ball striking a large cannon ball will recoil with some loss of energy, not necessarily large. The same billiard ball striking another of the same mass can effectively be stopped, that is, give up almost all of its kinetic energy, if the angle of encounter is right. The mass of the hydrogen atom is essentially the same as that of a neutron, so hydrogen is the most effective element of all for slowing neutrons and of critical importance to the shield designer. The other light elements are effective to varying degrees (the original Fermi pile used graphite as a neutron moderator) but hydrogen is best of all.

A neutron does not necessarily have to be slowed to the thermal range to be absorbed, but the chances of this happening are generally greater at the lower end of the energy spectrum. The absorption process itself and the sequence of events following absorption are critically dependent on the energy-level pattern of the struck nucleus. Absorption cross sections can thus vary wildly between isotopes of the same element. A gamma ray striking a gadolinium atom sees very little difference between ^{157}Gd and ^{158}Gd. A thermal neutron on the other hand is about 100,000 times more likely to be captured by the former as compared to the latter isotope. Differences in nuclear energy-level patterns also cause dramatic differences in plots of absorption cross section versus incident neutron energy for the individual elements and isotopes. Such a log-log plot for boron decreases rapidly and essentially linearly over the 0.01 eV to 0.1 MeV range whereas a similar graph for ^{238}U shows a series of very sharp absorption peaks in the "resonance" region between 1 and 400 eV. The difference of just a few electron volts in this "epithermal neutron" energy range can cause factors of hundreds in the ^{238}U absorption cross section.

A naturally occurring element will of course have an absorption cross section versus neutron energy pattern that is a composite of its mix of isotopes, and such a curve can show peaks and troughs. Iron for example exhibits very low minima at certain neutron energies. This can result in neutron streaming. A straight iron structural bar from one side to the other of a shield could thus permit excessive leakage at that point similar to gamma streaming through a duct. Substitution of a stainless-steel bar would help in this situation since the troughs in the iron cross section curve are compensated for to some degree by the peaks of the nickel in the stainless steel.[3d]

The mechanisms of neutron absorption have been briefly surveyed, but the question of the results has been only briefly considered in connection with inelastic- and elastic-scattering processes. In the case of very heavy nuclei the compound nucleus formed by absorption of a neutron can actually break up (fission). A neutron of the right energy taken into the nuclei of certain light-element isotopes such as ^{10}B can cause an alpha particle to be ejected, the (n,α) reaction. There are other situations where (n,p) or $(n,2n)$ reactions can occur. By far the most common pattern is however (n,γ) where a neutron is captured to form a new isotope of the target element. Some of the added energy in the system is simultaneously emitted as a "prompt gamma" (or a series of gammas). The new isotopes formed by "n-gamma" reaction with stable nuclei are frequently radioactive and measurable, a fact that is the basis of the technique of activation analysis.

The n-gamma reaction is a significant problem in designing shields against neutrons since their capture can occur anywhere in the shield. The resulting prompt gammas are of course similarly distributed in point of origin and add to the gamma protection problem. Boron-10 is sometimes incorporated into the inner layers of a shield since the alphas produced by the (n,α) reaction are readily stopped and neutrons that otherwise would give rise to prompt gammas are eliminated early in the game.

It will be seen that the designer faces problems when devising protection against neutrons, particularly when the source emits particles having a wide range of energies such as from a reactor. The first aim is normally to reduce the neutron energies as rapidly as possible to below the inelastic-scattering threshold, then to utilize materials to optimize elastic scattering and particle capture. These requirements are not entirely compatible, so there is a strong continuing interest in composite shields, that is, a layer of one material placed in front of another of a different type, or even a series of such layers. Calculations for such composite shields can become extremely complex. Further information is given by Glasstone and Sesonski,[54] Selph,[4c] Aronson and Klahr,[3d] Goldstein,[48] Schaeffer,[46] Courtney,[45] etc. Reliable cross sections for the different types of neutron reactions with matter are of course of critical importance to shielding calculations. Way and Carver[58] have published *A Directory to Neutron Cross-Section Data*. The most widely used printed compilations of the data themselves are probably those issued by Brookhaven National Laboratory,[59] the home of the National Neutron Cross Section Center (NNCSC). Similar centers exist in Great Britain, France, Russia and within the IAEA. Much computer-retrievable information is on tapes obtainable through these organizations.

4.5 SHIELDING: MATERIALS AND CONSTRUCTION

The discussion in the last three sections makes it apparent that no single shielding material is optimum for all situations. Light-element shields are best for betas and hydrogenous materials are the first choice for moderating and stopping neutrons. Heavy-element shields are superior for gammas and have a role in reducing initial neutron energies. The problem of these conflicting demands becomes particularly acute when shielding against an intense source of mixed radiations such as the core of a nuclear reactor where a severe economic penalty can result if the shield is substantially overdesigned and an even worse problem if the error is the other way. Hundreds of different substances have been considered and tested as shields and the "Engineering Compendium on Shielding[4]" devotes an entire volume to the subject. Structural adaptability, resistance to radiation and to high temperatures, physical bulk in some cases, and cost all have to be taken into account.

Shielding against a pure source of x- or gamma-rays is reasonably straightforward and certain materials—lead, steel, and concrete—are the usual choices. Since stopping gammas is primarily a matter of material density, lead is very effective and often serves as the standard to which the efficiency of other materials is compared. Lead has many advantages besides its high density (11.35). The metal is easily cast (melting point, 327°C), extruded, rolled and welded, and techniques have been worked out for machining. It can be obtained as slabs, pigs, sheets, powder, wool, shot and brick. It has good corrosion resistance and in the form of bricks allows the quick assembly and dismantling of shielding walls. The disadvantages are higher initial cost (partially offset by eventual salvage value); its softness, which can lead to mechanical damage if given rough handling; and its relative weakness as a structural material, particularly at higher temperatures. Lead taken into the body can cause central nervous system disorders and brain damage so casting and machining operations must be carried out with appropriate safeguards.

Lead is probably most frequently used in the form of straight-sided rectangular bricks, usually 2 in. in thickness, 3 or 4 in. in breadth and 6 or 8 in. in length. The flat sides however mean that straight-through cracks can occur in a stacked wall and lead to gamma leakage, particularly if the surfaces have been distorted by much use. The British have used lead extensively in construction of their hot cells and were the first[4a] to design bricks having mating surfaces in order to eliminate straight-through cracks. A variety of similar designs has since been developed by others. Figure 4.10 shows a version with curved surfaces that was produced in this country. Extrusion techniques were used for fabrication in this case which allows establishment of closer dimensional tolerances (cast lead shrinks about 4% on cooling) and eliminates the air bubbles that can occur if casting is not carefully controlled. The softness of pure lead brick can be substantially decreased by alloying with a small amount of antimony. The Brinell hardness is increased by a factor of 4 at the expense of a small ($\approx$ 2%) loss in density.[60]

The fact that lead can be obtained in so many different forms has generated many ingenious applications. Lead shot can be poured into cracks, as can the liquid metal, or lead wool can be pounded into small openings. Canvas blankets filled with lead wool or small shot can be wrapped around tanks, pipes or containers and lead sheet utilized in the same manner if only a limited amount

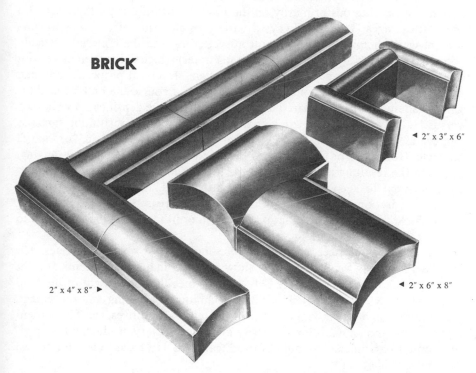

Figure 4.10 Lead-shielding bricks shaped to eliminate gamma leakage. (From Ref. 59, Courtesy N. L. Industries, Hightstown, N. J.)

of shielding is required. Special forms can be cast for fitting around valves and other control devices. Lead or lead plus boron have been incorporated into polyethylene for special applications.[4d] The metal has even been prepared as a handworkable putty that finds various uses in medical radiology and industrial radiography. The putty has about one third the shielding efficiency for gammas as does pure lead.[61]

Steel has not been used as extensively for shielding purposes as has lead, but its superior structural properties and lower cost can make it the material of choice in certain situations. Steel is of course used extensively in reinforcing concrete shields, and many hot cells are lined with stainless steel, not as shielding but as a means of obtaining a relatively corrosion-free and decontaminable smooth surface. Shielding walls entirely of steel are somewhat rare, but the material finds numerous applications in shielding doors, sample storage units, viewing window frames, penetrations, etc. Iron is reasonably effective in slowing fast neutrons by inelastic scattering, but the capture gammas formed by the eventual absorption of slow neutrons are hard—up to 10 MeV. Reinforcing steel in a concrete neutron shield accordingly must be placed with care. The problem of possible neutron streaming in such iron or steel components has been discussed previously.

Concrete is usually chosen as the shielding material for large, relatively permanent installations primarily on the basis of cost, although there are also other advantages. There is always a variety of elements of different weights present which helps in terms of neutron scattering, and the material normally contains 0.5%–1.0% hydrogen by weight (as water) which helps even more with the attenuation problem. Concrete has excellent structural properties and can be formed in the field. The main disadvantage is of course the bulk needed for gamma attenuation as compared to iron or lead (relative densities 2.4 versus 7.5 versus 11.3). This drawback can be partially overcome by substituting sands or aggregates prepared from heavy mineral ores for those used in "ordinary" concrete. Iron scrap or shot can also be added to the concrete during mixing, although the resulting mixes are hard to handle and require special techniques for placement to keep the metal evenly distributed. Minerals having bound water of hydration can be used if it is desired to increase the hydrogen content of the shield for neutron attenuation, or boron or lithium compounds added to reduce the capture gamma problem.

Table 4.3 summarizes information for some of these special concretes. The chemical formulas for the additives show only the major constituents and the final concrete densities given are only typical since the unit weight of the final shield will vary to some extent with the techniques of mixing and placement, the composition of the mix, the source of the mineral additives, etc. The serpentine concrete shown has a lower density than the ordinary product, but serpentine contains an unusual amount of fixed water ($\approx$ 12%) and thus is of interest for neutron shielding.[4e]

These special concretes are of course more costly than the ordinary variety and require experienced operators for concocting and placing a workable mix-

Table 4.3 Special Shielding Concretes

Material	Additives		Final Shield	
	Basic Formula		Density	lb/ft³
(Ordinary)	—		2.4	150
Serpentine	$MgO\text{-}SiO_2$		2.1	130
Limonite	$2Fe_2O_3 \cdot 3H_2O$		3.1	190
Barytes	$BaSO_4$		3.5	220
Ilmenite	$FeO\text{-}TiO_2$		3.5	220
Magnetite	Fe_3O_4		3.7	230
Hematite	Fe_2O_3		4.0	250
Ferrophosphorus	Fe phosphides		4.8	300
Iron shot	Fe		6.0	375

Source: Reference 4e.

ture. The planning of a large installation should thus be done carefully since ordinary concrete can be used in many places at a substantial savings and the heavy material reserved for those areas expected to receive greater amounts of radiation. The top 4 ft of a 12-ft-high hot-cell wall might be of ordinary concrete and a heavier type used only for the lower portions to give maximum protection to the operators at the working positions. Walls, floors and ceilings can also be made somewhat thinner if they are largely inaccessible or in tightly controlled access areas. The outer walls of the facility shown in Figure 4.2 are 4 ft in thickness for example, but the shielding doors to the individual cells are only 3 ft and the ceiling is 40 in. in depth.

Both ordinary and heavy concrete blocks can be cast and used for construction of less permanent shielding. Ring[4a] shows some quite elaborate brick designs for this purpose. Mating shapes should be used to avoid radiation leakage problems.

Concrete is a highly heterogeneous mixture of elements and it is probably safe to say that no two batches ever come out exactly the same. This fact does not simplify life for the designer or the engineer in the field, but the great importance of concrete to industry as a whole and to the nuclear portion in particular has generated an overwhelming amount of information concerning the material.[4e,3e,42,62,63] The American Nuclear Society has produced a standard,[64] NRC has issued a Regulatory Guide[65] and the Radiation Information Shielding Information Center prepares correlation reviews.[66] A type of reverse situation can occur where the natural radioactivity of the concrete is itself the problem. A Berkeley group[67] made a careful search for low background mix materials in connection with the construction of a low-level counting facility, for example. An interesting program was carried out[68] at Brookhaven National Laboratory to develop concrete-polymer materials as a means of introducing more hydrogen and for better retention of water for neutron attenuation. The list of such references could be extended indefinitely.

A shielding material that should be mentioned, primarily for gammas, is depleted uranium. Enrichment of natural uranium by concentration and removal of the ^{235}U has accumulated large stocks of very nearly pure ^{238}U. Uranium metal has a very high density (18.7) and thus is of great interest when it is desired to keep the thickness of shielding at a minimum, such as in the construction of shipping containers. At the other end of the scale is soil—a lot of it may be needed because of its low density ($\approx$ 1.5 dry) but there is a lot of it around and will provide good shielding if utilized in sufficient depth.[4g] Applications are chiefly in burying accelerator rings, waste tanks and other very large structures where provision of adequate above-ground shielding would be prohibitively expensive.

There are some situations, such as in handling (α, n) or spontaneous-fission sources, where the shielding problem is essentially all that of providing protection against neutrons, although some gammas will also generally be present from the source and the possible creation of capture gammas is always a factor. The goal in this case is generally to use a shielding material having as high a hydrogen content as possible. Water is 11.1% H by weight, roughly 6.7 × 10^{22} atoms of hydrogen/cm^3. Some plastics and elastomers are comparable or even higher[3e]:

Material	Density	Formula	Hydrogen density (10^{22} atoms/cm^3)
Butyl rubber	0.96	$(C_4H_8)n$	7.90
Lucite	1.19	$(C_5H_8O_2)_n$	5.73
Nylon	1.11	$(C_{12}H_{22}O_2N_2)_n$	6.56
Polystyrene	1.05	$(C_8H_8)_n$	4.86
Polythene	0.93	$(CH_2)_n$	8.02
Natural rubber	0.93	$(C_5H_8)_n$	6.54

Paraffin wax is a mixture of solid hydrocarbons, chiefly of the methane series, obtained from petroleum and having a density of about 0.9. It has been used extensively for neutron shielding but of course has the drawbacks of being low melting and readily combustible. Another high-H material is water-extended polyester (WEP).[69] Unsaturated polyester resins emulsify readily with water and when properly catalyzed the mixture hardens to a plasterlike substance that can hold up to 65% H_2O by weight and still remain a solid.

Pure water has been used both for hot-cell viewing windows (tanks with glass plates at opposing ends) or for shielding walls in the form of water-filled steel tanks. Such tanks, 12 in. in depth, have been used at Lawrence Livermore Laboratory to enclose a standard alpha containment box,[70] and 4-ft tanks filled with crushed magnetite ore flooded with borated water employed for a higher-level facility where gamma protection was also important.[71] The viewing windows in this last structure were a composite type, being a water tank on the hot side and slabs of shielding glass towards the operator.

Hydrogen, while the most efficient element for neutron attenuation, has a 2.2-MeV capture gamma which of course can be generated anywhere in a shielding tank as the neutrons are absorbed. This problem can be reduced by dissolving compounds of certain light elements in the water. The most useful of these elements are boron (for its ^{10}B) and lithium (for ^6Li). The indicated isotopes are among the few where the (n,α) reaction is overwhelmingly more probable than (n,γ). With ^{10}B however, the reaction product is an excited state of ^7Li which decays essentially instantaneously with emission of a 0.47-MeV gamma. Nevertheless, addition of a soluble boron compound to the shielding water effectively replaces some of the 2.2-MeV hydrogen capture gamma with emissions of lower energy. Lithium as an additive is in some ways even better since its (n,α) product is very weakly radiating tritium. Here one problem is that of finding a soluble lithium salt whose accompanying anion will not in turn lead to a new capture gamma hazard. The (n,α) reaction for ^6Li also has a lower cross section and the isotope makes up only 7.5% of natural lithium as compared to 20% ^{10}B in natural boron.

Boron and lithium have found many applications in shielding materials for special applications. Boral is an aluminum-clad alloy of boron carbide and aluminum. The material is of rather low structural strength but can be prepared in sheets, plates, tubes, and cylinders. Boron carbide dispersed in graphite was used as part of the shielding in both the British Dounreay and the American EBR-II reactors.[4f] Various boron-polyethylene and lithium-polyethylene specialty products are commercially available as are polyethylene bricks and slabs containing elements having unusually high capture cross sections for thermal neutrons such as cadmium and some of the rare earths. Boro-silicones, borated clay, and so on, are also available.[4e,61] Such materials are finding many uses where good stopping power for neutrons is wanted without having to pay a weight penalty, such as in design of shipping casks for isotopic and spontaneous-fission neutron sources.

The use of ^{252}Cf as such a source is in particular becoming steadily more widespread. The problems of shielding the isotope have of course received attention.[72-75] A similar sort of specialized problem is that of attenuating the 14.7-MeV particles from $^3H(d, n)^4He$ generators.[76-78] Still a third is in handling large quantities of plutonium and americium, during which subtle hazards can occur due to the (α, n) reaction.[79] Gamma facilities are also unique in that they generally handle only one isotope, usually ^{60}Co or ^{137}Cs, but in very large quantities. Arnold[80] considered both isotopes and their shielding along with many others as power-source possibilities, and Ross[81] gives more detail for ^{60}Co.

References have been previously given to the primary sources of shielding data, that is, tabulated attentuation constants, buildup factors, etc. Information is also available in more quickly accessible formats such as graphs and nomograms. Moteff [82] prepared a widely reproduced graph of gamma energy versus tenth-value layers for a number of materials. His curves are for the narrow-beam configuration so should be used conservatively in other applications. ICRP 21[30] gives a series of graphs showing the transmission of the radiations from some of the more commonly used radioisotopes through ordinary concrete, steel, lead and uranium. Steigelmann[31] published a similar set of curves, including a broader set of isotopes and with water also added as another possible shielding material. His report is now somewhat hard to come by, but some of his curves (for the dose delivered by the unshielded isotopes as a function of distance) have been reproduced in the Radiation Shielding Compendium.[4h] The appendixes of IAEA Safety Series No. 38[43] and No. 2[83] contain shielding information in graphical form. Arnold's report[80] on potential isotopic power sources contains much shielding information for the nuclides on his candidate list. Nomograms for quick calculation of shielding needs appear to be less

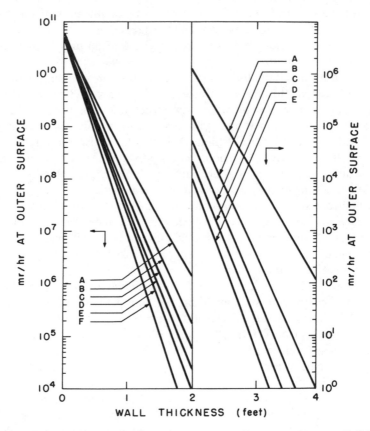

Figure 4.11 Shielding against 1-MeV gammas: curve A, ordinary concrete; curve B, 2.50-density ZnBr₂ solution; curve C, Barytes concrete; curve D, magnetite concrete; curve E, ilmenite concrete; curve F, Ferrophosphorous concrete.

prevalent than formerly. A number of these useful devices were published in Nucleonics, particularly over the 1955-1958 (volumes 13-16) period. An even earlier collection[84] of shielding nomograms is still useful.

The designers and operators of general-use shielded facilities are left with a problem. The radioactive materials with which they will be concerned are likely to be of various shapes, emitting gammas or particles over a spectrum of energies and quite frequently distributed as a number of sources from front to back of the working area. The only answer is to take the most pessimistic approach—assume that the maximum quantity of radioactivity anticipated for handling is concentrated as a point source close to the shielding wall. A representative energy value must then also be assumed. Figure 4.11 shows an example, calculated to aid in making a choice between various heavy concrete shielding walls. A 1-MeV megacurie gamma source placed 10 cm from the inner surface of the wall was assumed.

4.6 VIEWING AND LIGHTING

"Viewing" in hot laboratory terminology is concerned with the problem of seeing through or around shielding and of course is closely connected to the question of hot-cell lighting. Three of the techniques that have been developed—submerged viewing, closed-circuit television, and periscopes—will only be briefly considered since they are of limited value to the bench scientist.

Viewing has been discussed by Monk, Ferguson and Uecker,[3g] Ferguson, Doe and Goertz[3a] and by Ring,[4a] all of which papers give some consideration to cell lighting. Transparent shielding materials (windows) are reviewed in some detail by Jahn.[4i]

Submerged viewing has been used primarily in connection with handling of irradiated reactor fuel elements in storage pools and the underwater loading and unloading of large shipping casks. The handling devices used are of course massive, but the need for occasional finer manipulation has caused the development of special long-handled tools and examination devices which are handled by operators from above the pool surface. The water supplies surprisingly good viewing as well as shielding if care is taken to maintain the water quality. The approach is relatively inexpensive and admirably suited for simple operations, but the limitations are obvious.

Closed-circuit television would at first thought appear to be a useful approach to viewing of bench-scale hot-cell operations, but the technique has never caught on for such applications. The probable reasons for this are the poor degree of obtainable resolution and the fact that the image seen is essentially two dimensional unless "stereo" features are purchased at considerable added expense. Closed-circuit TV is however used to a considerable extent in reactor fuel reprocessing, waste tank management and accelerator operations where it is frequently necessary to monitor or to appraise situations in areas otherwise completely inaccessible to any other type of viewing. These are generally high-activity regions, which creates problems for the instrument designer who must concern himself with the working life of his optics and electronics in such an environment. A portable (4.6-Kg) TV camera that will take up to 10^7 rads before malfunctioning has been used extensively as an area monitor at the CERN accelerator in Switzerland.[86] Closed-circuit TV has some applications in ordinary hot-cell operations such as looking down into sample storage pits, monitoring restricted-entry transfer areas, etc. Small TV cameras also serve as the eyes for some of the robot manipulators discussed in the next section.

Periscopes have unique advantages for certain special situations such as in the examination of irradiated metallurgical samples, fuel elements and reactor hardware. They are far superior to other viewing devices if close details must be achieved through very thick shields. They are also the only device that can be used with instruments requiring "eye presence" in the cell, such as microscopes or optical micrometers. The chief drawbacks of periscopes are that the observer is closely confined to an eyepiece, that the field of view is limited (usually 50° or less), and that unless expensive modifications are made only one person can view at a time and will see essentially only a two-dimensional image. Being confined to the eyepiece position makes it difficult for the operator to simultaneously use manipulators and becomes very tiring to the eye if the procedure is a long one.

Standard optical glass darkens very quickly in high-radiation fields, so periscope components must either be shielded or specially compounded glass[87a] used if the instrument is to remain in such a field for any length of time. Cerium-stabilized optical glass offers some resistance to darkening or special optics may be cast from styrene or acrylic resins.

The barrel of the periscope must pass through the hot-cell wall in some manner and thus introduces a potential breach of the shielding. The horizontal penetration is generally made at a height several feet above the operator's heads with a minumum of two right-angled bends in the instrument, one down to the work level in the cell, and one in the external leg down to the operator's eye level. Supplementary heavy-element shielding at the entrance and exit points on either side of the wall may suffice for avoiding excessive gamma streaming. In other cases it may be necessary to add additional reflecting mirrors within the barrel so that the light path is shunted through the tube in such a way as to allow placement of lead or other shielding in the barrel itself. Other complications include the need for removal of the instrument for occasional maintenance or use at another working position, and in the case of controlled-atmosphere cells, to be able to accomplish this without affecting the cell atmosphere. Cell periscope systems can thus become quite elaborate.[88]

The best general discussions of periscopes and their applications in hot cells are probably still the older Refs.[3a] and [3f] cited above.

Simple reflecting systems find their use only in situations where the viewing problem itself is relatively simple. Such a system may be composed of a single mirror set to reflect a view around the corner or over the top of a wall of lead bricks in a hood, or may be made up of a series of mirrors set to propagate a reflection through a labyrinth in a shielding wall. The method is simple and reliable, but not very satisfactory if the required manipulations are complex. The angle of view is that subtended by the shield opening at the observer's point of view, and since this must usually be kept small because of gamma scattering, most installations are such that the static angle of view of the interior if of the order of 10° to 20° (obviously less true for a single 45°-angle overhead mirror). The angle of vision can be improved by use of a scanning mirror within the shield, but the shifting reflection makes it difficult for the operator to maintain his orientation, a problem for most people with only a single fixed mirror. The image is perceived at a distance equal to the developed optical length, which is generally two to five times the straight-line distance. Detail and depth perception are accordingly severely impaired. These limitations have restricted the use of simple reflective systems in more permanent hot-cell facilities, although they are still useful in connection with shields constructed for short-term operations with low amounts of activity or for a general view of a large area such as the back of a block of cells.

Windows, when they can be used, are without question the most satisfactory of the viewing devices since more elements of normal vision are retained than with any other technique. The field of view can approach 170°, unimpaired binocular vision is obtained at a minimum optical distance, nearly normal operator-equipment orientation exists during manipulation and, if proper cell lighting is chosen, color perception is good. The chief disadvantage, particularly for very thick windows having some intrinsic color, is that some light distortion and attenuation occur, leading to some loss of resolution. For ordinary viewing with the unaided eye, most of the resolution loss occurs because of chromatic aberration. This is most noticeable for viewing distances equivalent to a few thicknesses of the window or less, although color fringes may be observable at greater distances. If a telescope is used to obtain magnification through a thick window, spherical aberration may also become a problem.

Viewing windows are essentially holes in the shielding wall that are filled with glass or a high-density liquid, or even with pure water as mentioned under neutron shielding. Glass is the usual choice. The manufacture of shielding glass is a painstaking and expensive process and the market for the product rather small. Both Corning Glass and Pittsburgh Glass produced shielding glasses at one time, but, as far as can be ascertained, do not do so at present. Schott Optical Glass[87] and Nuclear Pacific[89] are apparently the only primary domestic suppliers at present. (Nuclear Pacific now operates Penberthy Instruments, one of the first firms in the business and developers of the first high-density glass—"Hi-D.") Overseas suppliers are located in Great Britain, West Germany, Japan and France, and the Nuclear News Buyer's Guide[90] lists a number of window designers and fabricators. The Schott and Nuclear Pacific brochures cited above and the articles by Ferguson and his associates and by Jahn list many of the physical-chemical properties of shielding glasses.

Glass materials specifically designed for shielding are largely a development of the period since 1949. Ordinary window (soda-lime) glass is about 20%

sodium oxide, 5% calcium oxide, and 75% silica. Plate glass is a thicker, more carefully made version. The glass manufacturers are rather shy about publishing details of their specific formulations for increasing the density of the product for shielding uses. This apparently is accomplished by substitution of barium or lead oxide for varying fractions of the soda-lime while at the same time reducing the silica content. (Penberthy's Hi-D contains 75% lead metal by weight.) A series of glasses of varying densities can be made as indicated in Table 4.4. Some of the earlier Corning and Pittsburgh Glass products also had densities in the intermediate 2.5 to 3.3 range.

An increase in glass density means an increase in gamma-shielding ability. Ideally, the window should give the same protection as the adjacent walls, and ordinary plate glass is a good density match to ordinary concrete. Polished plate glass is thus usually used as a cover plate on the cold side of windows where there is no need for a nonbrowning glass. Plate glass is available in thicknesses up to $1\frac{1}{4}$ in. (laminates up to 5–6 in.), and in sizes up to 74 × 148 in., although the maximum size that can be tempered is somewhat less. If the shielding wall is one of the dense concretes, the 2.5-density glass will not suffice by itself, so the inner portion of the window will usually be composed of slabs of some of the heavier nonbrowning materials. If more shielding is required to obtain a match to the walls, a high-density lead silicate glass is added to the mix. Such a composite window is shown in side view in Figure 4.12 for a 3.5-density concrete wall. (A nonbrowning 2.53 glass is the cover plate on the hot side.) In a lead wall, even a lead glass window will have to be built out on one or both faces to approximate a shielding match and such designs are frequently

Table 4.4   Shielding Window Glasses

Glass Type	Density	η^a	HVL[b]	TVL[b]
Schott				
RS 253	2.53	1.52	1.9	6.4
RS 323	3.23	1.58	1.5	4.9
RS 360	3.60	1.62	1.3	4.4
RS 420	4.20	1.69	1.2	3.8
RS 520	5.20	1.80	0.9	3.0
(Oil)	0.86	1.47	6.5	22
Nuclear Pacific				
Hi-D	6.20	1.98	0.8	2.7

Source: References 87 and 89.

[a] Refractive index.

[b] In inches against ^{60}Co without buildup.

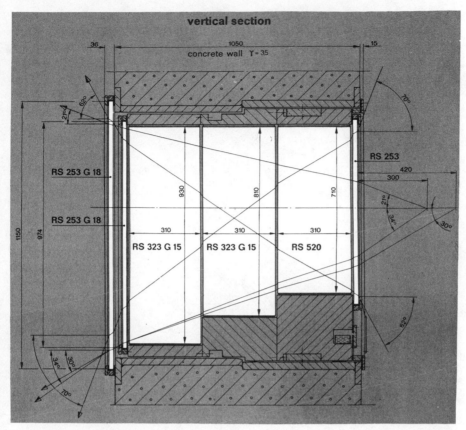

Figure 4.12 Composite all-glass shielding window. (From Ref. 87b, Courtesy Schott Optical Glass, Inc., Duryea, Pa.)

seen. Most cells, however, are only intermittently operated with the total amount of radioactivity that they are rated to accept, so the window can be designed to undermatch the walls if a portable shield of heavy glass is available that can be suspended on the operator's side of the window (if the wall is structurally able to take the weight) during times when the cell is loaded to capacity. Such planned undermatching of the windows is economically worthwhile only if there are enough windows in the complex to permit one portable shield to serve a number of different working positions.

The nonbrowning glasses mentioned above are obtained by addition of cerium oxide to the formulation in quantities up to about 2.5% by weight. (Higher amounts produce little additional benefit, and there is actually a negative effect for amounts over 6%.[91]) Unstabilized glasses start to show some coloration at doses between 10^3 and 10^4 rads. A yellow to brown color results after 10^5 rads, and at 10^6 rads the glass is almost opaque. (The pattern varies to some degree with the nature and the energy of the radiation and with the rate at which the dose is delivered. Interestingly enough, pure quartz is much more

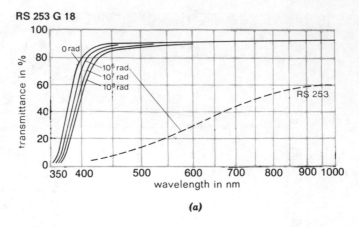

(a)

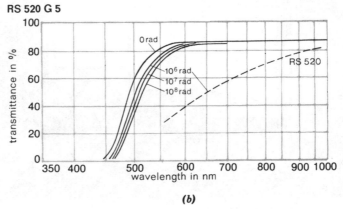

(b)

Figure 4.13 Effect of ^{60}Co gamma irradiation on light transmission of two shielding windows: (a) low-density, stabilized with 1.8% CeO_2; (b) high-density, stabilized with 0.5% CeO_2. (Courtesy Schott Optical Glass, Inc., Duryea, Pa.)

resistant.) Addition of the cerium provides a remarkable degree of protection as seen in Figure 4.13 which shows the effect on the light transmission of 10^6 to 10^8 rads of ^{60}Co radiation on two Schott glasses. Each of the graphs also shows (dotted lines) the effect of a 10^6-rad dose to the same glass with no cerium added. (In the Schott identification system the number following the RS is the glass density multiplied by 100, the number following the G, the cerium oxide content multiplied by 10.)

Radiation coloration in glass is apparently due to the formation of structural deformations which can act as traps for displaced electrons, thus forming "color centers." The phenomenon is complex, being affected by impurities, temperature, level of ambient light, rate of dose delivery, etc.[93,94] The color gradually fades after the irradiation is stopped, the rate at which this bleaching occurs again being affected by many variables. The fading is hastened by increased temperatures and generally by exposure to artificial lighting or daylight. The lead-containing glasses bleach more rapidly than the ordinary types. Cerium as an antibrowning agent is effective because the equilibrium between the Ce(III) and Ce(IV) oxidation states is such as to act as a buffer in absorbing some of the radiation energy.[95,96]

An intrinsic color can be seen in almost any glass when it is in the form of a thick slab. The color of lead glasses increases progressively with higher lead oxide content, this effect being intensified with cerium stabilized glasses. Even with no lead present, higher cerium content means more color. This can be seen in Figure 4.14 by comparing the curves for RS 253, RS 253 G 18, and RS 253 G 25 where the cerium oxide varies from zero to 2.5%. The glasses represented by the two bottom curves presumably differ primarily in their lead oxide content, resulting in a marked difference in the amount of (sodium-) light transmitted. The very-high-density glasses, particularly if cerium stabilized, can thus be used only in restricted thicknesses, their most common application being in lead structures where wall thicknesses seldom exceed 8 in. for structural support reasons. The high-density glasses are of course also used in multislab windows of greater thickness as a means of adjusting the overall density, as in Figure 4.2. In such a case the heavy glass is on the coldest side of the array of slabs so that cerium stabilization will not be necessary.

It might be noted in passing that TV-picture-tube glasses are cerium stabilized to reduce color changes caused by the electron beam generating the picture.

Another action of radiation on glass has its disconcerting side. The release and trapping of electrons can result in a buildup of electrical charge within a thick window, and a massive discharge can then be inadvertently brought

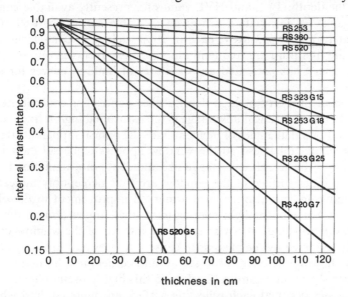

Figure 4.14   Internal light transmission of various shielding glasses to sodium light. (Courtesy Schott Optical Glass, Inc., Duryea, Pa.)

about by any shock that cracks the surface of the glass. A bright flash of light is seen and a dendritic pattern of pulverization forms in the body of the glass. The number of occurrences of this rather unnerving phenomenon worldwide in working hot cells over the last twenty-five years is somewhere around ten, and, as would be expected, generally in the most highly exposed glass slab nearest to the hot side of the window. There has naturally been research on causes and modes of prevention—one paper in the Seventh Hot Lab Proceedings, three in the Eleventh, and one in the Eighteenth (see Refs. 2g, 2k, and 2r, respectively). Electrical measurements show, that as with color fading, the accumulated charge dissipates over varying periods of time after cessation of exposure. The presence of cerium in the glass seems to aid in inhibiting the charge buildup and several specimens of stabilized commercial glasses resisted discharge even when subjected to very strenuous experimental conditions.[96] There is some evidence that slabs below a threshold thickness will dissipate the charge rapidly enough to avoid the charge buildup. Culler[97] states that the use of a 1-in. thick hot-side stabilized glass cover plate is sufficient to eliminate damaging levels of buildup in the other glass components of the window.

The shielding ability of viewing window glasses is generally presented only in terms of gamma radiation. The previously cited brochures from Schott and Nuclear Pacific give graphs of tenth-value layers versus gamma energy for their respective products, and the former includes an extensive set of buildup curves which is also reproduced in the review by Jahn.[4i] The quantitative effect of including buildup in the shielding calculations will necessarily depend on the thickness of the actual window and the energy of the gamma, but in most cases the effect is to require somewhere between one-half and two tenth-value layers added to the depth. TVL and HVL values for presently available commercial glasses against ^{60}Co radiation are given in Table 4.4. (Cobalt-60 emits two gammas at 1.17 and 1.33 MeV, each in 100% yield. For convenience, tabular values in the literature are frequently more or less averaged out at 1.25 or 1.3 MeV.) Lindner[98] has presented an earlier set of buildup curves for shielding glasses.

A physical property of high-density glass (and of a concentrated $ZnBr_2$ solution) that is of practical advantage is the associated high refractive index, η (Table 4.4). For small angles of incident light the optical (apparent) thickness is $1/\eta$ of the actual thickness, so the window appears to be substantially more shallow than is the actual case. Image size and depth discrimination for objects close to the inner surface of the window are thus perceptibly increased. The high refractive indices also produce a marked improvement in the observer's field of view. Some of the light coming in at small angles around the edges of the window that would otherwise strike the interior of the window opening is sufficiently refracted by the glass as to cause it to pass out through the window face, thus allowing the operator to see around corners to a limited extent. The angled lines shown in Figure 4.12 relate to this highly useful effect.

Light losses occur at each reflecting surface encountered, and while these are relatively small individually and reducible by special surface leaching and coating techniques, can become significant in a thick window made up of a number of slabs. One method of reducing these losses is by filling the spaces between the slabs with a transparent oil having roughly the same refractive index as the glass itself (the "oil" of Table 4.4). An oil-filled window of this type is a "wet" installation; with no oil, it is "dry." Even the best of the oils will be subject to radiation damage which usually causes darkening or the formation of deposits, so the choice of the filling liquid must be carefully made and provisions installed for draining and replacement when necessary. A relatively thin window will normally be dry installed. Some of these have a low flow of dry nitrogen between slabs to prevent atmospheric moisture and CO_2 from etching a haze into the glass.

Shielding window design thus has its complications.[99] High-density glass is costly, so an effort must be made to minimize the quantity needed without jeopardizing the needs of the operator. Possible gamma streaming through cracks around the outer surfaces means that the windows are usually stepped, as in Figure 4.12, and supplementary shielding may be necessary. Thought must also be given to dismantlement of the window if repairs become necessary or if oil darkening or deposits become so severe as to require a cleaning of the surfaces of the glass slabs. The larger and more complex windows generally consist of a steel casing for the glass itself, which casing is then fitted into a steel liner cast into the wall. This presents another problem if neutrons as well as gammas are present since care must be taken to avoid straight runs of metal through the wall that could serve as neutron pipes. Design and engineering of a large shielding window requires experience and careful thought. Some of these aspects are considered in more detail in previously cited references.

Direct viewing through a shield is of course not always so complex if only a restricted area need be seen, if the wall is relatively thin and if the radiation levels are not high. A single solid slab of glass may suffice as a window in such a situation. Such "peepholes" having the dimensions of standard lead bricks are commercially available,[89] and can be readily incorporated into a lead shielding wall. The British make extensive use of cylindrical glass castings to act as viewing ports into their lead brick cells. The operator must of course stand close to such small windows which limits the type of manipulation that can be used.

Filling the window cavity with a transparent liquid is markedly less costly than filling it with glass. Distilled water may be adequate when neutrons are more of a problem than gammas.[100] If much gamma activity must be handled, the thickness of pure water required soon makes its use impractical and recourse must be had to liquids of higher specific gravity. Doe[101,4i] was the original investigator in this field and examined a number of liquids heavy enough to be considered for shielding window use. Lead acetate (specific gravity, 1.5) has been used for shielding against x-radiation and is quite effective for this purpose. Acetylene tetrabromide, a tin tetrabromide-tin tetrachloride mixture, a solution of barium and mercuric bromides, zinc chloride, methylene bromide, cadmium borotungstate, thallium formate-malonate, etc., were all considered as possible window materials but were discarded because of their toxicity, cost or instability. The material chosen as most suitable for development was zinc bromide. This salt is very readily soluble and solutions containing as much as 78.3% dry weight can be prepared, having a density of 2.54 (a very good match for an ordinary concrete wall) and a refractive index of 1.56. Doe considered the manufacture and properties of zinc bromide solution at some length and included a set of purchase specifications for the material. A highly purified salt is needed, available in quantity. Currently this offers a practical problem since the only known domestic supplier, The Michigan Chemical Company (now known as Velsicol), no longer produces the material, although there are indications that there are sources in Israel, Japan, and Great Britain.

Ionizing radiation will cause oxidation of zinc bromide solution and cause the release of free bromine. This can be controlled by addition of 4 to 40 oz of hydroxylamine per 100 gal of solution, although the protective effect is eventually lost when the preferentially oxidizable hydroxylamine is depleted. Where unusually high levels of radiation are anticipated the zinc bromide solution should be protected by enough glass shielding to reduce the radiation at the interface to 5×10^4 R/hr or less since bubbling may occur at exposures above that level. Such protective shielding was incorporated in the window shown in Figure 4.15.

An interesting approach has been tried in placing the zinc bromide solution directly in a mild steel tank with no surface coating, but under a blanket of argon gas.[102] The associated facility was in use only for a few years but the window apparently gave no problems during that time. A "gamma-neu" window has been suggested.[103]

The Argonne Wing M cave complex contains a total of forty large zinc bromide windows that have been in use for over seventeen years. When in good condition these windows are very satisfactory to use, but there have been problems, primarily resulting from solution penetration of the coating originally used on the tank interior (seven successive coats of Amercoat paint). Subsequent reaction with the steel of the tank has caused darkening and precipitate formation at one time or another in practically all of the windows. Some organics may have also been dissolved from gasketing materials to add to the coloration. A white, cloudy suspension, possibly zinc hydroxide, has also formed in a few cases.

The zinc bromide solution can be quite satisfactorily cleaned by filtration, although this is a laborious and sticky process. Ceramic filters are best, and bubbling of oxygen or ozone through the liquid during filtration helps in eliminating color. A recent approach [104] has been to dilute the solution with an equal part of distilled water. 30% H_2O_2 is then added to precipitate the iron. The solution is then readily filtered but the clear filtrate must then be concentrated by heating until a density of about 2.5 is again reached.

The solution in a deteriorated window is pumped out for clarification and the original tank coating is removed down to the bare metal. Experiments were tried with cast polyethylene liners fitted inside the tanks as replacements for paint coatings. Some of the windows so rebuilt have given no further problems,

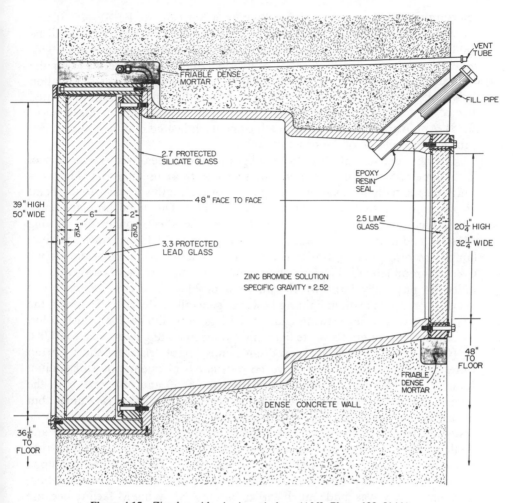

Figure 4.15 Zinc bromide viewing window. (ANL Photo 120-5144.)

but in some cases the liners have cracked. Current practice is to use an epoxy-resin coating (again seven successive coats). Two windows so treated experimentally in 1969 are still in good shape, the same being true of other units subsequently receiving the same treatment.

One possible viewing (at least light-transmitting) device that would seem to have some potential for hot-cell work is fiber optics, but apparently there have been few applications. Radiation damage to fiber optics has received considerable attention[105] in connection with instrumentation of space vehicles where exposure to cosmic rays and other space radiation is a problem. These studies have included cerium stabilization of fiber-optics systems.

4.6.1 Cell Lighting The depth of the window, its overall light-transmission quality and the nature of the work to be performed are the primary factors in determining the design lighting level within a hot cell. At HFEF-N[106] it was decided that an "as-viewed" light level (that seen by the operator) should be 100 fc 3 ft back from the window interior face at the working-table level. This value was then divided by a series of factors. Allowances were made for the decrease in lamp output during its working life (0.8), for lamp-glass radiation darkening (0.7), for window darkening due to radiation (0.8 at an integrated flux of 2×10^9 R) and for the measured light transmission of the window (0.38). Application of these various adjustments indicated that the initial in-cell light level should be of the order of 600 fc.

The next decision needed after the light level has been fixed is with respect to the type of light to use. Sodium lamps such as those used on some highways emit an essentially monochromatic light at a wavelength very near to the point of optimum sensitivity of the human eye (550 nm). The monochromatic emission largely eliminates the problem of chromatic aberration in the viewing window, and thus is highly desirable where viewing of detail is critical. Sodium lamps are also more operationally efficient in terms of output (lumens/w). Color discrimination is however eliminated, the lamps are more costly and the units are physically large which can be a problem in a crowded cell. The decision on use of sodium lighting however generally reduces to balancing the advantages of sharper viewing against the loss of ability to distinguish color.

Mercury lamps emit a white light, have rated operating lives of up to 24,000 hr (versus perhaps 8000 hr for sodium lamps), have relatively short startup times (5–10 min), are smaller, have no restrictions of orientation (the sodium units cannot be suspended downward from the base) and of course have the reverse of the sodium-lamp pattern in that color discrimination is retained, but at sacrifice of some clarity in detail perception. Both types of lamps have been tested for operating characteristics under prolonged gamma irradiation[107] and while there were minor differences, commercially available units of either type were evaluated as quite suitable for hot-cell work. HFEF-N decided upon mercury lamps almost solely on the basis of retention of color discrimination. The same choice was made much earlier for the ANL Wing M facility because of the importance of color judgments to the radiochemist.

Incandescent- or fluorescent-lighting systems are of course also possibilities and are sometimes used in smaller cells. These types of lighting are however normally employed in larger cells only as lower-light-level systems for use when the cells are entered for cleanup or equipment building operations, or for interior lighting of in-cell containment boxes. Small incandescent spotlights are sometimes used to light particularly important areas. Incandescent light is however a jumble of wavelengths which exacerbates the chromatic aberration difficulty in thick windows.

HFEF-N uses a reserve quartz-iodine lamp at each working station, connected to the emergency generator which comes on automtically if other power is lost. Light is thus available for safe shutdown of work in emergency situations.

Both the mercury and the sodium lamps are available in bulbs of various wattages. A frequently used size is the 400-W unit. These lamps must be located within the cell so that they do not shine directly into the eyes of any of the operators at the viewing windows. For this reason, and to place the maximum amount of light on the working area, the lamps are usually installed on the wall above or at the sides of the inner face of the windows. If the cell is such as to have opposing windows, the lamps must be shaded or louvred.

The major remaining difficulty is that of maintenance and repair. Even the best of the lamps eventually burns out and must be replaced, and the associated wiring and hardware do not last forever. The solutions to these problems depend on the nature of the operation and its effect on the frequency with which the cell can be entered after start up. A radiochemical experiment may be over in a matter of days, or months at worst. The expectation then is that the cell will be entered for cleanup before a new program is started and that any deficiencies that have developed in the lighting system can be remedied at that time. Other programs, particularly in controlled-atmosphere cells, envisage re-entry only in terms of years or even decades, which adds a new dimension. Repairs have to be made remotely.

One solution is to install steel-lined penetrations into the wall at time of construction. A plug designed to contain the light, its conduit, and other hardware is then built to fit into the liner and can be pulled out of the wall when repairs or replacements are necessary. The same problems of supplementary shielding and contamination control are then met as in the case of periscopes, mechanical manipulators, transfer drawers or any other device initially requiring an open hole between the inside of the cell and its exterior. Ring[4a] reproduces several Oak Ridge designs for this type of light-fixture plug. The other approach is to build portable units that can be handled and plugged into outlets in the cell with available manipulation. The light is placed in a unit having suitable protuberances for aiding manipulator handling and to enable hanging on preset wall brackets in the desired lighting positions. Failure of an individual lighting unit then becomes a matter of always having a replacement ready for movement into and installation in the cell, with the replaced unit being removed as contaminated equipment.

The high level of light intensity required even in a hot cell with only moderately thick shielding walls adds considerably to the heat-disposal problem for the ventilation engineer, particularly in controlled-atmosphere or recirculated-air cells. One concept in the very early days was to have no ventilation at all in order to confine the radioactivity. The problem of handling the resulting build-up of heat very quickly scotched that approach.

Ferguson et al.[3a] have given a discussion of cell lighting in addition to the references already cited.

4.7 MANIPULATION

In the hot laboratory vocabulary manipulation is concerned with mechanical devices (sometimes powered) controlled by a human operator and used to move or manipulate radioactive objects without direct contact. The definition as given is very broad and could cover all the ground between a pair of tweez-ers and the largest of the reactor-core removal mechanisms. The present discus-sion will however be largely restricted to the use of devices permitting an experimenter on one side of a shield to carry out laboratory-type operations with radioactive materials on the other side. Such laboratory procedures are of course designed around the human animal and his manipulative assets—pri-marily the senses, arms and hands under the direction of the brain. The engi-neers are making steady progress towards effective mechanical duplication of parts of this unique combination, and have actually improved upon it in some ways—ability to lift enormous weights to considerable heights, wrists that make full 360° turns, and resistance to high-radiation fields, high temperatures and lack of oxygen are examples. Fortunately, replacement of the human brain component still appears to be a very considerable distance in the future.

Goertz, the outstanding pioneer in manipulator development for nuclear applications, pointed out[108] some years ago that even a reasonable approxima-tion of the movements of a human arm and hand requires a minimum of seven degrees of freedom as shown in Figure 4.16. A beaker on a bench can be moved left or right (x motion), toward or away from the operator y motion), or directly up and down (z motion). By using the wrist as a pivot, the beaker may be moved in an arc in a horizontal plane (azimuthal motion), raised in an arc in a vertical plane (elevation), or tipped as in pouring (twist motion). The act of holding or releasing the beaker requires fingers having a squeeze or grab action. If a more complicated maneuver must be performed, such as turning a crank, it can be seen that this is possible with a manipulator limited to these degrees of freedom only if the direction in which the force is applied can be smoothly and easily reversed. Goertz has termed devices having a minimum of the described seven degrees of freedom as general purpose manipulators. In many cases the operations do not require all seven motions, so more limited devices have their place, particularly if special equipment for the working area is devised with the limitations of the available manipulator in mind.

The simplest of these devices is the scissors-action tong whose advantages and limitations are familiar to every chemist. These are useful in handling materials of just sufficient activity to make it desirable to avoid any near direct contact. Additional advantage can be taken of the attenuating effect of distance by fitting the tong to a long flexible or rigid tube carrying a mechanical linkage to allow the operator to cause a set motion to occur at the operating end by pressing a trigger, pushing a button, turning a knob, etc., at his end. The most common operating-end motion is that of squeezing so that objects can be grasped and moved, but other actions are obtainable. Stang[109] describes a num-ber of interchangeable heads that can be fitted to the end of a tong, for exam-ple, lifting hooks, forceps, saw, screwdriver, tweezers, parallel-jaw pliers or sheet-metal scissors. He also pictures a device using a spring-loaded expand-able steel band that can be drawn tight or loosened for handling beakers or test tubes.

The simple tongs are usually used in the open in situations where distance alone is sufficient to provide exposure protection. Simple tong-type motions are

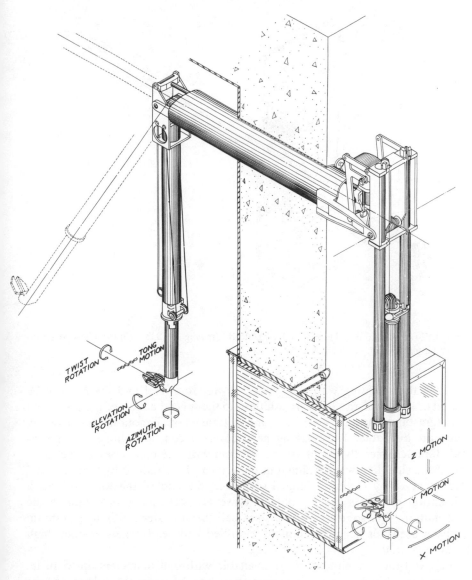

Figure 4.16 The Argonne Model 8 master-slave manipulator. (ANL Photo 106–2063.)

however generally adequate for performing operations in the restricted-volume enclosures of the Concentrate and Confine system if the equipment is specially designed for tong use. When shielding is needed, facilities under this philosophy essentially become shielded gloveboxes as seen in Figure 3.3. "Ball-and tong" manipulators were developed to meet the problem of introducing the tong without violating the integrity of the shielding as seen in the same figure. The handle of the tong is passed through a metal sphere set in a close-fitting

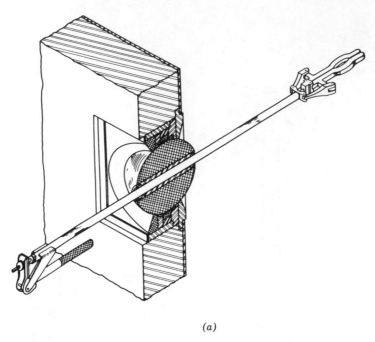

(a)

Figure 4.17 (a) Ball-and-socket manipulator (ORNL drawing A-13337; (b) cylinder-disk ("Castle") manipulator. (ORNL Photo 15296.)

socket in the shielding wall. The sphere is free to swivel and the rod can be rotated or slid back and forth. In addition to the special action of the tong head itself (grasping, cutting, etc.) x, y and z translational motions and twisting action are available. The working head covers a conical volume with an included angle generally about 70° to 90° and with the cone apex at the inner side of the shielding wall. A different pattern can be obtained by "articulated" tongs, that is, by having a hinged section at the end of the tong rod. Such articulation generates other problems however, particularly in obtaining the twist motion. The sphere can be a solid ball (usually steel, although uranium has been used), or it may be hollow and filled with lead pellets or other high density material.

Ball-and-tong manipulators are available with ball diameters up to 10 in. (for an 8-in. shielding wall), but frictional drag between the sphere and its socket makes the device difficult to operate smoothly if the ball diameter exceeds 4 to 6 in. [Figure 4.17(a).] This in effect limits the use of the manipulator to wall thicknesses of not much more than 4 in. if the manipulations are at all complex. The "Castle" manipulator [Figure 4.17(b)] improves on this limitation and allows reasonably good handling through walls in the 8-10-in. range. Discussion of both devices is given by Ring,[4a] Ferguson et al.,[3a] and Stang.[109]

The limitations of pure tong handling and the availability of master-slave types of equipment have undoubtedly caused a decreased amount of interest in

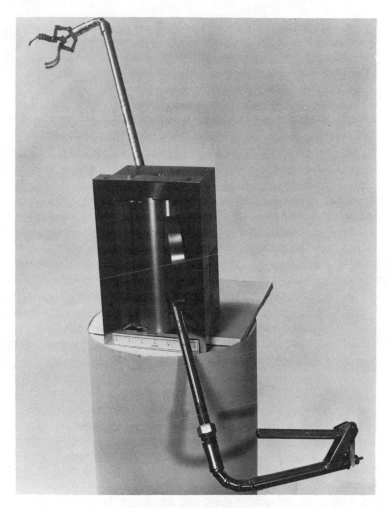

Figure 4.17   (Continued)

ball-and-tong and Castle manipulator use in recent years. This is also probably true of laboratory-scale rectilinear manipulators. In this type a single arm is suspended on a bridge which rides on overhead tracks at the inner and back faces of the cell along its length. Movement along the tracks gives the x motion and back and forth across the bridge gives the y motion. The z motion is obtained by means of a telescoping tube or similar device in the arm. Azimuthal, elevational, and pouring motions become possible if the arm is given a wrist. A control box having a separate switch for each of the motions is operated by the experimenter from his position before the viewing windows. Manipulations must be made slowly in order to avoid accidents, and it is difficult to design a unit where two "hands" can be used to work together. Simple rectilinear manipulators are less suited for laboratory than for heavy-duty work, a large

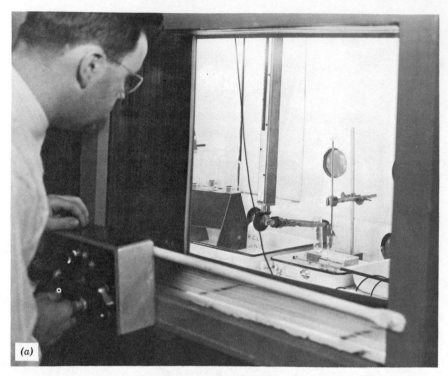

Figure 4.18 (*a*) Simple rectilinear manipulator. (ANL Photo 107-2207.) (*b*) Polar manipulator and alternate "hands." (From Ref. 110, Courtesy Gesellschaft für Kernforschung mbH, Karlsruhe.) (*c*) Self-propelled manipulator. (Courtesy PaR Systems Corp., St. Paul, Minn.) (*d*) Truck-mounted rescue manipulator. (Also from Ref. 110.)

bridge crane being an example. Such cranes are routinely used in reactor and fuel-reprocessing facilities to accomplish remarkable feats of remotely connecting, maintaining, repairing and disconnecting large items of equipment.

The simple rectilinear manipulator has only a "wrist." If an "elbow" is added, more complicated motions are possible and if a "shoulder" joint is also added, the manipulator gains motion capabilities much nearer to those of the human arm and can be considered to operate in a polar rather than a rectilinear set of coordinates. If the arm is mounted on a vehicle capable of independent motion through the cell and if television viewers are mounted on the same unit, the combination begins to approach the true robot category. The basic arm can also be wall mounted, made operable up and down a column, hung on the end of a boom or suspended from a bridge crane. In each case a console having separate control for each of the possible movements is needed, so some practice is required before an operator becomes truly expert.

Figure 4.18(*a*) shows a simple rectilinear manipulator in a chemistry cell being operated by means of a portable control box from the outside. Figure 4.18*b*) pictures an unmounted polar manipulator arm of a type used by the Nuclear Emergency Brigade of the Federal Republic of Germany,[110] a group

(b)

Figure 4.18 (Continued)

formed and equipped for rescue and emergency work in case of accidents involving radioactivity. The interest in this photograph is partially in the wide variety of special heads shown for attachment to the basic arm—power drills and saws, wrenches, cutters, grabs, etc. Figure 18(c) shows a movable unit built in this country by PaR Systems Corporation.[111] The working arm (with two elbows in this case) moves up and down or rotates on a center column carried on a tank-track-mounted self-propelled vehicle. Adjustable TV cameras furnish the robot his "eyes." Figure 4.18*(d) is a photograph of another piece of equipment available to the German Emergency Brigade, presented as an indication of the range of polar manipulator sizes that have now been designed and built.

The French have also built a number of interesting self-propelled manipulators, using wheels rather than tank treads as in the PaR version. Their "Virgule"[112] has two working arms operable in master-slave fashion, can climb stairs or clamber over low obstacles and can draw its wheels together for passage through narrow openings. The "MA 23"[113] is the latest French entry and was developed with underseas as well as nuclear work in mind. Its two slave arms are removable so that either normal or heavy-duty types can be used. The electrical system is being developed so that the machine can be operated by direct control, in a tape recording and playback mode, or by computer control.

Figure 4.18   (Continued)

Development of equally elaborate machines is underway by a number of groups in this country and abroad, many of which activities are reported upon in the RSTD Proceedings, including a complete session devoted to manipulators in 1977.[114]

The type of manipulators just described are of course quite costly and are not very suitable for fine-detail manipulation, so their use as laboratory tools has been limited. Their best applications in nuclear work are probably for equipment installation and repair, for emergency use such as retrieving a spilled sample from the floor or a corner, loading or unloading shipping casks, cell decontamination, cell decommissioning, etc. Similar manipulators have also entered other fields—underseas work as indicated above, mines, and on space vehicles where their future use will undoubtedly grow when the day comes for assembling large structures in space.[115]

Another interesting pattern that appears to be at the point of rapid expansion is the use of robots in industry to perform the more dangerous, hot or

(d)

Figure 4.18   (Continued)

tedious jobs that hitherto have necessarily been performed by humans. Such machines have been under development since the early 1960s.[116] Industrial robots are now a 60 million dollar a year industry, predicted to grow to 3 billion by 1990.[117] These robots ("Unimate"[118] is probably the best known example) have to be extremely rugged and reasonable in cost, but in general are only required to carry out rather simple repetitive actions, but to do these rapidly and for extended periods of time. These requirements are in contrast to those in most nuclear facilities, particularly for laboratory operations, where complex actions must be carried out, but usually on almost a one-time-only basis and with time generally available for a more deliberate approach. The manipulative needs are thus quite different.

There is little question but that most of the general-purpose manipulators in use today for laboratory-scale hot laboratory operations are of the master-slave type diagrammed in Figure 4.16. Master-slaves can be either purely mechanical, as shown, or electro-mechanical. In the mechanical type the "orders" are given by the operator's movements and are passed from the control (master) arm to the operational (slave) arm by direct mechanical linkages—metal tapes, rotating force seals, steel cables, etc., with assistance from small electrical motors for some actions. In the second type, the orders are transmitted electrically, but in such a fashion that the slave-arm motion follows that of the master. Development of both types was largely pioneered by R. C. Goertz and his associates at Argonne National Laboratory, with a multitude of variants now being produced and used around the world. In this country two Minnesota firms[111,119] are the leading suppliers, both of long standing.

Figure 4.16 shows the standard ANL Model 8 manipulator. As in all similar machines, the mechanical linkages are carried though a penetration tube in the shielding wall set at a height of about 10 ft above the floor. Two manipulators are normally installed at a working position, one for the operator's right hand, the other for his left. The slave arm hangs at right angle in the cell, and the master arm is similarly suspended on the cold side of the viewing window. Each movement of the master arm is duplicated by a similar motion of the slave arm and squeeze action is initiated in the fingers of the slave hand by squeezing action on the control handles of the master. The widespread acceptance of the Model 8 is largely due to the fact that the transmitted motions follow those of the operator so closely, making its operation easy to learn, even for manipulations requiring considerable dexterity. The other substantial advantage is in having two hands working against or in assistance to each other on the same piece of work. The basic Model 8 arm is furnished with a two-fingered hand which acts so as to move the fingers together in a parallel fashion. These hands can be removed and replaced remotely by the use of special jigs, and a variety of replacements are available to perform specialized operations. Stang[109] describes a number of hand tools adapted for operation with the device.

If the object being manipulated is off to one side of the center of the working area or uncomfortably far back, manipulation becomes awkward if the master arm position must duplicate that of the slave arm exactly. Electrically driven "indexing" features have been designed for these problems. Operation of a switch on the master end can cause the slave arm to be moved sideways to the extent desired, or by use of a telescoping tube, to considerably extend movement in the y direction ("extended reach"). It is also convenient on occasion to be able to leave an object held in a fixed position in space when the operator releases the controls. "Side-indexing," "extended-reach" and "motion-stop" options are accordingly available commercially. The standard Model 8 will handle objects up 20 lb (9 kg) weight. Other versions are available for either "light-duty" (10 lb) and up to "rugged-duty" (100 lb) application.[119] An ultra small "Mini-Manip," which combines the ball-and-socket wall insert of the ball-and-tong manipulators with limited master-slave action can be purchased.[111] The device will handle 5-lb objects but can only be used in relatively thin shielding walls. Vertut and his associates[120] have made a study of the cell-space coverage afforded by the standard Model 8 and a number of its variants. The article shows interesting diagrams of a number of the latter.

The purely mechanical master-slave requires a wall penetration which introduces the potential for radiation streaming, contamination release and air leakage. Lead inserts around the linkages in the manipulator barrel can be used and supplemental shielding placed on the inner wall below the manipulator to eliminate all but scattered radiation at the tube inlet. "Booting" (or "socks") is the most common solution for the contamination problem. The entire slave arm down to the wrist is enclosed in a flexible rubber or plastic bag having one end sealed to the machine, the other to some sort of glove ring arrangement encircling the penetration opening on the wall. Installing or removing the boot can involve pull the manipulator back through the tube, always a wearisome business because of the contamination problem. Jenness[121] has described a two piece boot that permits remote replacement of only the lower half where failures are most likely to occur. Parsons and his associates[122] have reported on the more usual master-slave booting technique.

Controlled-atmosphere cells are a special problem any time a penetration is required. Special Model 8 manipulators have been developed[119] in which master-arm movements are coverted to pure rotations which are transmitted through double-rotary, dry inorganic face seals between the master and slave mechanisms. Leak rates through the seal are less than 2 $(cm^3/hr)/cm$ of water pressure differential. Another approach is with quadruple, nitrile rubber, oil-lubricated lip seals in the manipulator barrel to provide an essentially zero-leak rate.

The completely mechanical master-slave manipulators have several limitations. The constraints imposed by the mechanical linkages means that a somewhat limited volume of the cell is covered, although this can be improved by use of the extended-reach and side-indexing features. As pointed out above, gas-tight seals are needed in some situations. The force exerted on the work is ordinarily limited to that that the operator can exert on the controls (although special grips are available to give some powered amplification of the squeeze action), but usually the experimenter must furnish all of the power for carrying out the manipulations. These considerations have led to the development of the more elaborate (and considerably more expensive) electrical-mechanical manipulators in which direct linkage is replaced by electrical wiring or even radio or laser-beam communication between the master and slave arms.[123]

The strictly mechanical master-slaves provide a surprisingly good sense of force reflection or "feel" to the hands and arms of the operator. This feature is of obvious importance, particularly where the equipment handled may be of a fragile nature, as in chemical research. Replacement of the mechanical linkages loses this direct force reflection and necessitates the use of a separate bilateral, force-reflecting positional servomechanism with appropriate feedback for each of number of degrees of freedom for which the manipulator was designed.[124] Removal of the mechanical linkage allows the slave arm to be suspended on a bridge crane, boom, etc., to provide greater cell coverage. Force multiplication also becomes easier to engineer. While these are unquestionably very definite advantages, the higher cost and complexity of the electrical-mechanical manipulators require rather special situations for their justification. Most researchers on the laboratory scale find the workhorse Model 8 and its offspring to be quite satisfactory for their needs.

4.8 MATERIALS TRANSFER

Supplies, tools and equipment must be routinely transferred into cells, gloveboxes or containment boxes during operation as well as the active starting materials. Samples, waste and contaminated items must be brought out. The problems encountered in such transfers have been met before—strict control of operator exposure and spread of contamination, and, in the case of controlled-atmosphere enclosures, avoidance of dilution of the cover gas.

Many individual transfer devices for particular installations have been described in the Proceedings of the Hot Laboratory Group,[2] and there is discussion in Barton[125] (for gloveboxes), Ring,[4a] Ferguson et al.[3a] and Welsher.[126] Reviews specifically dealing with materials transfer techniques have been made by Schulte et al.[6a] end by Lefort.[127] These last two reports attempt to establish general classifications for transfer situations, and, while the nomenclature is different, the two evaluations are not too dissimilar.

Movement of clean material into a contaminated enclosure is generally not too difficult unless the item is so large as to require the opening of a shielding door or its equivalent. Much more usual is the everyday need for smaller items. Since this is basically a one-way traffic (the materials will probably end up as active waste) it is best to reserve one transfer mechanism exclusively for such cold items to give better overall control of the spread of contamination. (Such segregation may not always be possible as with many gloveboxes where a single radiation lock must serve for all transfers, hot or cold. Such a lock is seen on the back of the containment box pictured in Figure 4.3 The negative pressure within the box helps control contamination while materials are being moved through the two doors of the lock. The latter will nevertheless inevitably become contaminated and, if accessible, require periodic cleaning. A separate bag-out system for solid-waste disposal materially helps the situation.)

Cold transfers into a cell handling moderately low levels of activity can be a relatively simple matter, particularly if no alphas are present. A wall penetration of suitable size delivering to a receiving tray within the cell and normally closed by a plug or a movable slab of shielding at the cold end can be used. The penetration (port) is briefly opened and the item pushed through onto the tray with a rod or tongs. The penetration hole is sometimes slightly sloped downwards to expedite matters. Another simple device is the sphincter, one form of which is shown in Figure 4.19(a). A series of flexible disks, each having two slits cut perpendicularly to each other in the center, are held upright in the pass-through port. The item to be transferred (most conveniently in a can of standard size) is pushed through the array of discs, each of which closes itself as the passage is completed. The device shown is for a cell, but is also adaptable for some glovebox uses.

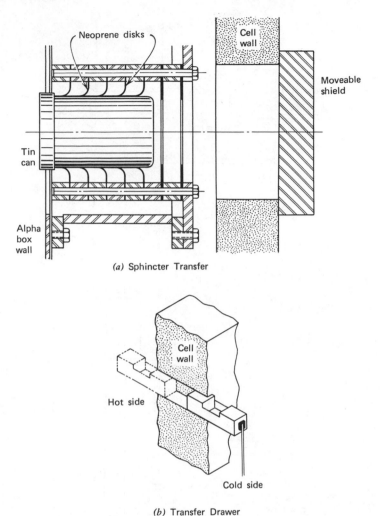

(a) Sphincter Transfer

(b) Transfer Drawer

Figure 4.19 Some materials-transfer devices. (From Ref. 6a, Courtesy Los Alamos Scientific Laboratory.)

The transfer devices must become more complicated if the radiation and potential contamination problems are more severe. Transfer drawers passing through the wall are sometimes used. In its simplest form, such a drawer is an open box fitting into a wall opening of the same areal cross section and riding on rails. Shielding is placed in each end of the drawer, the space in between being for holding the items to be transferred as shown in Figure 4.19(b). The dimensions are such that the shielding at one end of the box or the other always closes the wall opening so the radiation exposure is minimized. Contamination spread is more difficult to control, but if necessary delivery can be into a gloved box at the cold end. Transfer drawers are obviously more suitable for traffic in

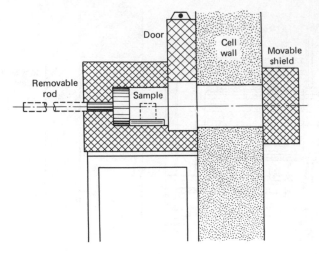

(c) Port—Cask System

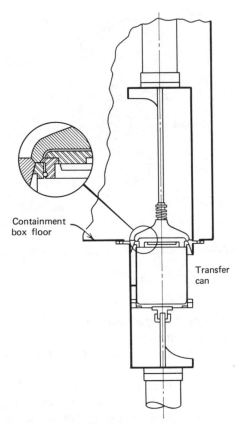

Containment
box floor

Transfer
can

(d) Double—Cover Device

Figure 4.19 (Continued)

either direction from the cell interior than are simple ports or sphincters if there is no other satisfactory means of bringing samples and contaminated small items out, but constant monitoring is required.

Moving radioactive materials received in shipping casks is a more difficult problem. If the cask is small enough, moving into and emptying the container in the cell itself may be the simplest approach, although a decontamination operation will be needed later. In some cases the cask is moved into a special cell or area (the "pot-unloading area" of Figure 4.2) and the exterior first protected from contamination as much as possible by shrouding with plastic sheeting. The cask is then opened remotely and the contents transferred to the working cell by a conveyor belt or cart, by placement in a transfer drawer if the unloading area is adjacent to the cell (transfer areas are frequently used for unloading), etc. (The Wing M transfers at ANL are accomplished by "mules," that is, small battery-driven, radio-controlled carts running on narrow-gauge tracks throughout the complex.[128] The analytical chemistry facility at Oak Ridge[16] is an in-line set of eight cells, one of which acts for unloading and as the sample distribution center. The shipping cask containing samples from the outside is wheeled on a transfer cart into the cell through a shielded radiation lock. The cask cover is removed and the cask raised by a pneumatic lift to a level convenient for the manipulators to remove its contents. These are placed in temporary storage racks to await distribution to the working cells, this last being accomplished by a conveyor which runs the length of the complex in front of and just below the viewing window.)

Bulky and high-activity items such as fuel assemblies or reactor-loop components require considerably more engineering than the relatively simple approaches hitherto described and are usually some modification of the cask-port system diagrammed in Figure 4.19(c). A specially designed cask mates with locking hardware against an opening into the cell wall, floor or ceiling. The "door" shown in the diagram is part of the cask and is pulled far enough out of the way to allow the contents to be "rodded" into the cell after the inner shield has been moved aside. The rod is removed during shipment. When the time comes for use it is passed through a hole in the cask bottom and fastened to a slab of movable shielding that can be pushed forward to slide the cask contents into the cell. There are numerous variations, some of which are discussed by Schulte el al.[6a]

Transfers into cells can also be made by conveyors passing through L-shaped penetrations or labyrinthes. Variations of the bagging technique are also used for inward- as well as outward-bound materials. These, even if clean, are often bagged-in (most frequently in glovebox work) since the procedure always leaves a positive contamination barrier in place after the bag is sealed off inside the enclosure. A type of Chinese-box approach can be used in moving contaminated materials out. The item is first bagged or canned in a primary container whose exterior is assumed to have become contaminated. This is then moved to a cleaner area where the primary container is sealed in a secondary. The process can be continued indefinitely and triple bagging is not unusual.

Bagging (Figure 3.12) for eliminating solid waste and contaminated items produces a neat, high-integrity, easily handled and positively sealed package. Use of the method is accordingly becoming more and more common, even for very large items.[6a] An alternative for glovebox and containment-box work is the "double-cover" technique, an example of which is shown in Figure 4.19(d), used with considerable success at Los Alamos.[129] A closed friction top can (original version very costly, but since found replaceable with one-gal paint cans[25],) is raised by a lift to fit into a seal on the bottom of the box. This seal encircles an opening that is covered by a lifting device as shown in the figure. When the can is firmly in place, the lifter pulls off the lid and moves it away, keeping its top surface covered and protected from contamination. The opened can is then filled from inside with samples, wastes or any other item to be transferred. The lifter then reseals the lid and the can is dropped down to a level accessible to the manipulators. The only exterior part of the can susceptible to contamination during the process is an annular ring of about 0.25 mm width around the top edge. This area can be taped over as a precaution before any further movements are made.

In a multicell complex with a single-cask loading-unloading area some method has to be provided for transfers within the facility itself, both for distribution of incoming samples and for the centralization and packaging of outgoing material. The Oak Ridge use of a conveyor and the ANL cart-rail system for this purpose have been mentioned. Transfer drawers or ports in the dividing walls of in-line cells may be used for passage of materials. If the whole facility is bridged by a rectilinear manipulator, movable blocks in the intervening walls may be pulled out of the way by an overhead crane to allow intercell transfers. Adaptation of toy electric trains to move small items through wall tunnels between locations has also been used. Solutions to material-transfer problems have thus varied, depending very much on local facilities and program. The simplest system that still provides adequate safety protection should be chosen. The movable bits of hardware should be kept to a minimum and arranged with all the vital mechanisms located where maintenance and repair with safety can be assured. Critical mechanical equipment access is probably best located in the transfer area, less desirably in the operating area, and should never be restricted to the enclosure interior itself unless it is clear that any breakdown can be remedied by remote means.

The relationship between the national governments and the nuclear industry and its suppliers in most overseas countries is somewhat more direct than in the United States. This is probably the reason that standardization of specifications and designs for transfer devices, manipulators, viewing windows, shielding wall components and other hot laboratory hardware appears to be more prevalent abroad. France,[130] West Germany[31] and Great Britain,[4a] for example, apparently have catalogues of standardized equipment that would allow construction of many facilities simply be ordering and assembling the right components. The U.S. situation is more informal, and while a number of governmental and professional groups previously cited are establishing standards, these are generally in terms of performance rather than detailed design.

The AEC for some years published the "Engineering Materials List"[132] which gave specifications, drawings and other design information for nuclear hardware.

Some types of operations require at least short-term storage of samples and other materials. The largest of the cells in the Oak Ridge analytical chemistry facility is reserved for this purpose, as an example. Storage needs, as for any laboratory, should be evaluated during the design stage, particularly if fissile materials are being handled with the attendant possibility of a nuclear criticality accident if storage is haphazard. Bringing materials out of a cell is a time-consuming affair so there is a natural tendency to allow items to accumulate after their period of usefulness is over. This is obviously not the best of policies for a number of reasons.

4.9 SERVICES (UTILITIES)

The spectrum of services called for in the design of an ordinary laboratory is fairly constant, but this is much less true for a hot cell since so much depends on the intended use of the facility. Some cells may require nothing more than electrical power delivered to the interior, others may need most of the utilities normally furnished to a laboratory. There are two general principles that should be followed, whatever the selection. The first is that all penetrations such as conduit or pipes should be offset by at least one diameter in passing through the wall in order to minimize radiation streaming. The second is that the primary controls for the services should all be on the exterior of the cell, preferably grouped around the viewing window within easy reach of the operator.

Figure 4.20(a) sketches the two-position face of a cell designed for chemical research.[133] The services at each window are identical. The piped utilities are brought up just below the viewing window where they are valved off. The electrical power is similarly brought to outlets at the lower sides of the window. A matching set of switched outlets further up at the sides are directly connected through conduit (bent in an S shape) to a third set of outlets in the cell interior on a shelf before the window, clearly visible to the operator and accessible to the manipulators. The double-outlet system at the face then allows placement of a rheostat or other control device into the line, or simple connection by a patch cord as shown in the drawing. The other utilities are handled in a similar fashion. The row of black dots below the window represents a series of $\frac{1}{2}$-in. stainless-steel-pipe penetrations that are again brought to takeoffs along the utility shelf in the interior. Valves or tubing connectors can be installed at that point, or the penetrations can be used for passage of flexible tubing for carrying reagent solutions. On the exterior, the break between the valved-off utilities and the penetration inlets again allows the incorporation of any required special control devices. Another set of penetrations is also furnished for connections to readouts and exterior controls from instruments being used in the cell. All of

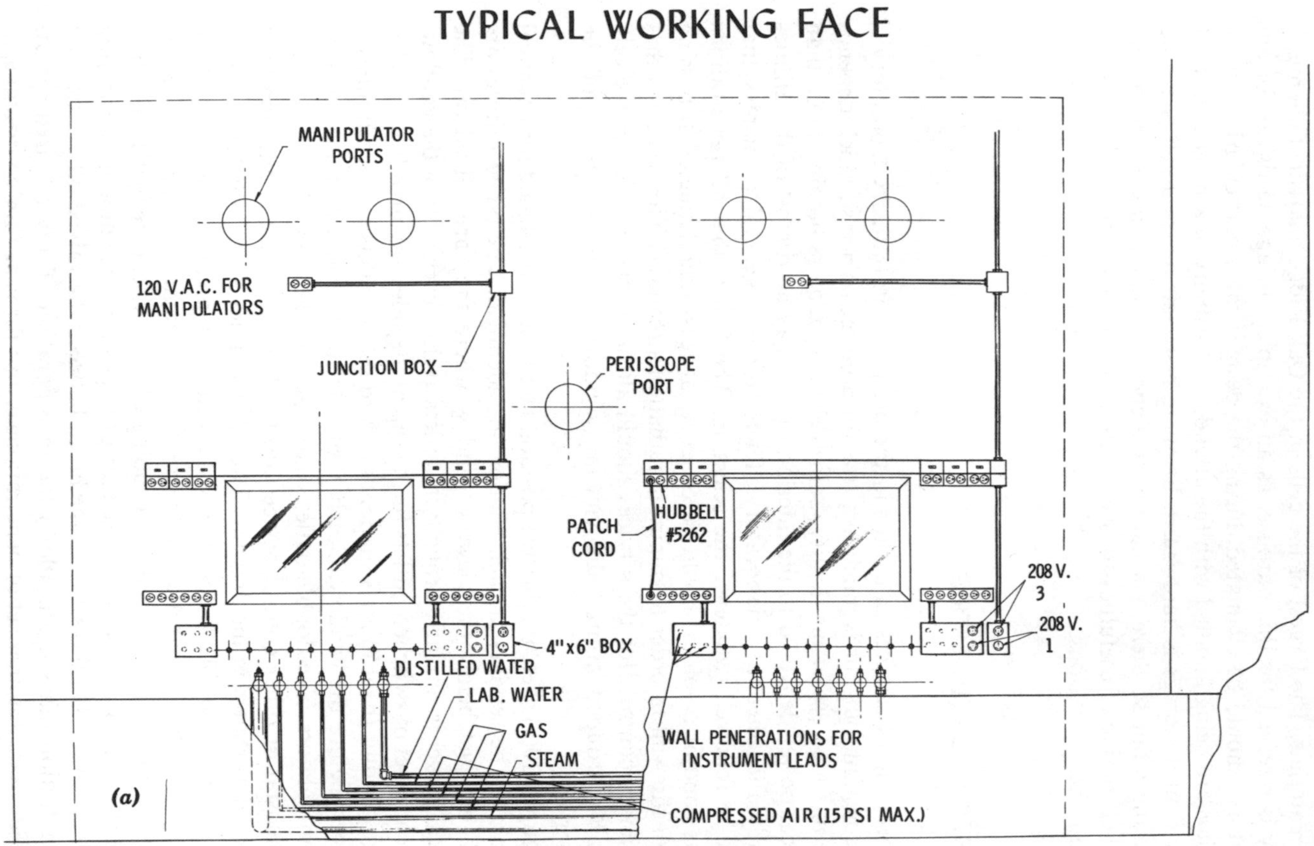

TYPICAL WORKING FACE

Figure 4.20 (a) Sketch of utilities distribution at cell operating station. (b) Photograph of cells in use. (ANL Photos 120–4191 and 308–682.)

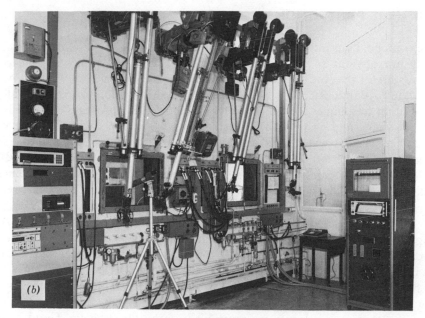

Figure 4.20   (Continued)

these various penetrations again follow an S-shaped pattern to avoid streaming problems and can be capped or plugged when not in use.

The three pipes marked 'gas' in Figure 4.20(a) are not for natural gas, a service that presents too much potential hazard for hot-cell use. The lines instead run to cylinder racks out of the way along the inner walls of the building at the sides of the operating area. Gases such as carbon dioxide, nitrogen, argon, etc., for special purposes can thus be utilized without having cylinders crowding the operating space at the windows. This system has worked very well. A steam line was also furnished in this particular installation with the thought that steam would be useful for preliminary cell decontamination during cleanup operations. The steam system has actually found very little use, certainly not enough to justify the cost of its installation. The distilled and laboratory water services shown are needed for a chemistry cell, but probably should be completely eliminated for facilities where the potential of nuclear criticality is a factor to be considered.

If utility penetrations are cast into a shielding wall, as is usually the case for concrete cells, an effort should be made during construction and before the piping and conduit are buried from sight to record their positions in space on 'as built' drawings. While in theory this information should be on the original building plans, on-the-spot modifications inevitably occur during construction. Boring of a new penetration hole in the wall may become necessary at some later date, in which case reasonably precise knowledge of the location of power, steam and compressed-air lines may help avoid unpleasant surprises.

4.10   FIRE PROTECTION

The complications introduced into a fire emergency by the presence of radioactivity are obvious so many precautions must be taken. Following the very serious fire at the Rocky Flats installation near Denver in 1969,[134,136] the AEC delegated Atlantic Richfield Hanford as the lead laboratory to reexamine fire-prevention techniques for hot laboratory facilities. This program continued for several years, and as with the Livermore study on filters with which it was closely associated was funded for a complex of laboratory, theoretical, and literature evaluation approaches. The final summary report appeared in 1974.[137] The Savannah River Laboratory, because of its heavy involvement with alpha handling, conducted a somewhat parallel program.[138,139] Other references not as directly pertinent are in the various publications of the NFPA.[140]

Fires are classified according to the material doing the burning: Class A—paper, wood, plastics, etc.; Class B—flammable solvents; Class C—electrical; and Class D—metal fires. Any or all of these types may occur in nuclear facilities, although the probabilities for each will vary with the program. Class D fires are a major concern in glovebox work with plutonium and other pyrophoric metals, but improbable in most radiochemical or nuclear research where paper, plastics and solvents are more likely to be the problem. Good housekeeping and other preventative measures if conscientiously applied can reduce the chance of a fire to almost zero. If one does occur, three critical items are needed in place: some means of almost instantaneous detection, a more-than-adequate supply of an appropriate extinguishing agent and a plan for informed and effective response to the emergency.

Various types of fire detectors are available, broadly classified as those responding to light, heat, pressure changes, and to products of combustion generated by the fire. (Hill[138] gives clear descriptions of the different mechanisms used in each type of device.) Some detectors are inherently more suitable for one type of fire than another—a smoke detector may or may not react to combustion of pure lower alcohols, for example. Other complications arise in the case of hot cells. Because of the accessibility problem, the detector must reset itself automatically after having acted, or be susceptible to remote resetting. A detector in any location should be periodically tested to be certain that it is in operating order, and if found defective, it should be replaced. These actions are difficult to accomplish remotely. Corrosive fumes and sometimes dust can be present in hot-cell atmospheres, and radiation damage to the wiring and the unit itself may reduce the working life far below normal. The detector must thus be exceptionally rugged and not prone to false alarms while at the same time being as simple in design as possible with a minimum of radiation-susceptible parts.

Both the Atlantic Richfield and Savannah River programs have made systematic surveys of commercially available detectors. Thermal detectors can respond when a fixed temperature is reached, be actuated when the rate of temperature increase attains a predetermined value, or be a combination fixed-temperature and rate-of-rise device. Photoelectric smoke detectors are of two types. In the first, smoke coming into a beam of light falling on a photoelectric cell causes enough attenuation to activate the detector. In the second type, the beam does not shine directly on the cell but is located so that smoke coming into the beam diffuses light onto the cell to generate the response. Both types will give a false alarm if dust or some solid object enters the beam. Ionization detectors contain a radioactive source which ionizes the air between the source and a collector plate, allowing a small and constant flow of electrical current. Visible or invisible products of combustion entering the detector chamber absorb some of the source radiation, causing a change in the current flow and activating the alarm system. (Ionization detectors are probably the worst suited for hot-cell use because of the above normal and variable amount of air ionization already present.) Surveillance detectors are usually either of the infrared- or the ultraviolet-light type. The former are customarily set to respond to infrared radiation in a narrow spectrum range, modulated at a frequency of 13 Hz, the frequency associated with flames. Ultraviolet detectors respond to light generated by flames at the other end of the spectrum.

Hill's conclusions[138] after completing his survey program were: "Thermal detectors are more reliable and suitable for remote locations where maintenance and testing is difficult or not feasible. The response rate of smoke detectors is similar to that of thermal devices, but significantly slower with solvent fires, especially the low molecular-weight alcohols. Light detectors respond faster to fires, but they require more maintenance and are more prone to false alarms. Products-of-combustion detectors (ionization type) lose sensitivity in a high radiation field." Lee[137] reached very similar conclusions.

Two different detector types can be used to give an extra degree of protection. When containment boxes are being used, a rate-of-rise unit can be placed in the top of the box, backed up by an infrared type peering through the viewing window at the box working area.[141]

Lee lists water, inert gases (including CO_2 and Halons), dry chemicals, foams, and steam as possible fire-extinguishing agents. Some of these are more suitable for use on certain types of fires (CO_2 may accelerate oxidation in metal fires, for example) and there are again complications introduced by the special circumstances of nuclear installations.

Water is normally available in sufficient quantity for fighting even large fires, but is not always 100% effective when metals or organic solvents are the combustibles. Water can also damage electrical and other equipment, a problem of more than ordinary seriousness in a hot cell because of the difficulty of making repairs, and will almost certainly spread contamination and leave a messy cleanup operation. Nuclear criticality possibilities must also be considered. Nevertheless, an automatic water-spray system may be the most practical alternative for very large cells or canyons if criticality is not a factor.

The chief problem with the more truly inert gases such as nitrogen or argon is that of ensuring that a large enough supply is always on hand for smothering the largest conceivable fire, although these agents could be considered for a small cell or a few gloveboxes. The gas must be carefully introduced into the enclosure in order to avoid sudden overpressurization, already a potential danger because of the increased temperatures and presence of combustion gases due to the fire. On the other hand, a sufficient concentration of the gas for blanketing the fire may be difficult to attain unless the normal exhaust from the enclosure can be quickly reduced, again introducing relative pressure problems.

Carbon dioxide also acts by blanketing a fire to exclude oxygen and is best for solvent and electrical fires and also quite effective for combustibles of the paper-plastic type. The problem of adequate supply again has to be considered. Such supplies are usually in the form of liquid CO_2 which will condense to form crystals when expanded as a gas into a cell or glovebox. These crystals interfere with visibility and can clog the exhaust filters, although these are not insurmountable difficulties.[142]

The Halons are halogenated low-molecular-weight aliphatics, the one having received the most attention being Halon 1301 (bromotrifluoromethane). Only 5 vol% in air is effective since the compound decomposes very rapidly, the resulting products then reacting chemically with intermediates involved in flame propagation. (Carbon dioxide must be at more than 45 vol% for equal results.) Some residual corrosive gases are produced, but in dilute form because of the relatively small amount of suppressant required.

A major advantage of gases for fire control is that they leave no mess of their own behind requiring subsequent cleanup. Dry chemicals do leave such residues to add to the decontamination problem and perhaps to clog the exhaust filters. Even equipment not directly involved in the fire may require remote disassembly and cleaning. Dry-chemical extinguishers have accordingly found little application in hot-cell work. They are more useful, at least as a backup system, in glovebox operations.

Certain organic materials when mixed with water and air and sprayed through special nozzles will produce long-lasting foams that are particularly effective in extinguishing Class A and Class B fires. The foam apparently acts both as a smothering agent and as a carrier of water for removal of heat. Lee[143] classifies the foams as high-expanson, low-expansion and aqueous-film forming, and has tested each type extensively. The difference between the first two is largely in the volume of foam produced per unit weight of material. The aqueous-film type forms a foam that will float on the surface of organic solvents and is also a very good wetting agent. The foams would seem to be most applicable in large enclosures where the use of inert gases would be impractical. The foams do have drawbacks. The high-expansion type is difficult to introduce into an enclosure without overpressurization unless the air required for mixing is taken from the cell itself, in which case the presence of combustion products may make a good foam difficult to produce. The foam, because of its light weight, may also be drawn into the filters. The low-expansion foams dry to a fluffy mass that is hard to remotely remove from walls, floors and equipment.The aqueous-film type may cause problems when the cleanup liquids are introduced into waste-processing systems.

Steam as a fire suppressant has been used in the past in naval vessels and factories where large quantities of steam were always available. Its application in hot cells and glovebox trains for firefighting is very questionable in view of the overpressurization, filter-clogging and contamination-spread possibilities.

The administrative organization and automatic devices for responding to a fire emergency will vary widely depending on circumstances, but obviously merit much advance planning and continuing attention. An important and difficult question arises because of the hazardous nature of the materials being handled, although this problem is not entirely unique to the nuclear field. In many cases, particularly during nonworking hours, a blaze may have to be dealt with by professional firefighters having only slight acquaintance, if any, with the facility and the location and dimensions of possible hazards from the radioactivity. There does not seem to be any easy answer to this problem other than extensive labeling, warning signs and perhaps a continuing program of two-way education between those responsible for a facility and those who might be called upon to meet an emergency in it.

REFERENCES

1 U.S. Department of Energy (DOE), "Western New York Nuclear Service Center, Companion Report." USDOE Report 28905-2 (1978); DOE, 100 Independence Ave. SW, Washington, D.C. 20545.

2 Proceedings of the 'Hot Lab Group.' (See text) The first eight meetings of this organization were during 1950-1960. The Proceedings (Refs. 2a-2h here) were published by a variety of sponsors, a complete listing of whom is generally given in each of the later volumes. The Proceedings since 1961 have been issued by the American Nuclear Society (Chap. 1, Ref. 12). Volumes 9-11 were titled 'Proceedings of the Xth Conference on Hot Laboratories and Equipment.' Subsequent volumes are 'Proceedings of the Xth Conference on Remote Systems Technology.' Volume 25 contains a cumulative Index for Vols. 1-21. (i) Vol. 9 (1961); (j) Vol. 10 (1962); (k) Vol. 11 (1963); (l) Vol. 12 (1964); (m) Vol. 13 (1965); (n) Vol. 14 (1966); (o) Vol. 15 (1967; (p) Vol. 16 (March 1969); (q) Vol. 17 (November 1969); (r) Vol. 18 (1970); (s) Vol. 19 (1971); (t) Vol. 20 (1972); (u) Vol. 21 (1973); (v) Vol. 22 (1974); (w) Vol. 23 (1975); (x) Vol. 24 (1976); (y) Vol. 25 (1977); (z) Vol. 26 (1978).

3 *Reactor Handbook, 2nd Edition.* Volume I, 'Materials.' C. R. Tipton, Ed. (1960); Vol IIIB, 'Shielding.' E. P. Blizard and L. S. Abbott, Eds., (1962); Vol. IV, 'Engineering.' S. McClain and J. H. Martens, Eds., (1964). Wiley Interscience, New York: (a) K. R. Ferguson, W. B. Doe and R. C. Goertz, 'Remote Handling of Radioactive Materials' Vol. IV, pp. 463-538; (b) U. Fano, C. D. Zerby and M. J. Berger, 'Gamma Ray Attenuation.' Vol. IIIB, pp. 102-107; (c) A. B. Chilton, 'Buildup Factors.' *Ibid.*, pp. 63-101. (d) R. Aronson and C. N. Klahr, 'Neutron Attenuation.' *Ibid.*, pp. 63-101; (e) H. E. Hungerford, 'The Nuclear, Physical, and Mechanical Properties of Shielding Materials.' Vol. I, pp. 1027-1101; (f) G. S. Monk, K. R. Ferguson and D.F. Uecker, 'Remote Viewing.' (This paper was in Vol. II lst Ed., of the Handbook and also given in USAEC Report AECD 3646 (1955).

4 R. G. Jaeger, Ed.-in-Chief, *Compendium on Radiation Shielding*, Vol. I 'Shielding Fundamentals and Methods.' (1968); Vol. II, 'Shielding Materials.' (1975); Vol. III, 'Shield Design and Engineering.' (1970). Springer-Verlag, Berlin: (a) F. Ring, Jr. and R. Smith, 'Design and Shielding of Hot Cells for Research.' Vol. III, pp. 95-134; (b) J. H. Hubbell and M. J. Berger, 'Attenuation Coefficients, Energy Absorption and Related Quantities.' Vol. I, pp. 167-184; (c) W. Selph, 'Neutron Attenuation, Interaction Processes.' Vol I, pp. 259-261; (d) W. W. Parkinson and O. Sisman, 'Homogeneous and Inhomogeneous Combinations; Plastics, Heavy Mate-

rials and Boron.' Vol. II, pp. 342-345; (e) H. E. Hungerford, A. Hönig, A. E. Desov, F. DuBois and H. J. Davis, 'Concrete, Mortar and Grouts.' Vol. II, pp. 75-270; (f) F. Åcker-heilm, 'Boron and Boron Componds.' Vol. II, pp. 336-342; (g) D. Spielberg, 'Soils (Nonhomogeneous Combinations).' Vol. II, pp. 345-368; (h) H.H. Eisenlohr, 'Air.' Vol. II, pp. 265-270; (i) W. Jahn, Silicate and Lead Glasses.' Vol. II, pp. 55-72; (j) W. B. Doe, 'Zinc Bromide Solution.' Vol. II, pp. 72-74.

5   Chapter 3, Ref 2a

6   European Nuclear Energy Agency (ENEA), 'Working Methods in High Activity Hot Laboratories.' *Proceedings of the Symposium at Grenoble, France, June 15-18, 1965.* Organization for Economic Cooperation and Development, Paris (1965). In two volumes: (a) J. W. Schulte, P. J. Peterson, M. T. Wilson, G. J. Deily and E. S. Fleischer, 'Methods Used in the United States for Transferring Materials through Radiation Barriers.' pp. 523-544.

7   IAEA (Chap. 1, Ref. 2), Radiation Safety in Hot Facilities.' *Proceedings of the Symposium at Saclay, France, October 13-17, 1969.* IAEA Proc. Ser. IAEA-SM-125/, STI/PUB/238 (1970): (a) I. K. Shvetsov, V. N. Kosyakov and D. V. Morozov, 'Design of Transplutonium Laboratory with Circular Arrangement of Cells.' pp. 535-545.

8   Chapter 3, Ref. 21

9   Chapter 3, Ref. 2.

10 A. Valentin, L. Hayet, and J. L. Faugére, 'The Radiometallurgy Laboratory at Fontenay-aux-Roses, Equipment and Startup of Alpha, Beta, Gamma Cells.' (Ref. 2o above), pp. 206-216.

11 A. R. Irvine, A. L. Lots and A. R. Olsen, 'The Thorium-Uranium Recycle Facility.' (Ref. 2m above), pp. 19-24.

12 J. Hesson, M. J. Feldman and L. Burris, 'Description and Proposed Operation of the Fuel Cycle Facility for the Second Experimental Breeder Reactor (EBR II).' USAEC Report ANL 6605 (1963); (also see Ref. 2t above, pp. 261-270).

13 N. J. Swanson, J. R. White and J. P. Bacca, 'Startup Configuration of the Hot Fuel Examination Facility/North (HFEF/N)' (Ref. 2w above), pp. 109-126.

14 V. W. Eldred and K. Saddington, 'The Post Irradiation Examination Facilities at Windscale Works, U.K.A.E.A.' (Ref. 2i above), pp. 264-288

15 Chapter 3, Ref. 27.

16 L. T. Corbin, W. R. Winsbro, C. E. Lamb and M. T. Kelley, 'Design and Construction of ORNL High-Radiation Level Analytical Laboratory.' (Ref. 2k above), pp. 3-10.

17 A. L. Coogler, G. J. Deily and R. J. Hale, 'Evolution of the High Level Caves at Savannah River Laboratory' (Ref. 2m above), pp. 102-114.

18 A. B. Ritchie, 'Safety Experience and the Control of Hazards in Hot Cell Operations at Harwell.' (Ref. 2o above), pp. 167-178.

19 P. F. Moore and J. D. Allen, 'Los Alamos Radiochemistry Hot Cells.' (Ref. 2k above), pp. 11-16.

20 W. E. Bost, Comp., 'Hot Laboratories, an Annotated Bibliography.' USAEC Report TID 3545 (Rev. 1) (1965).

21 Anon, 'Appendix A, Survey of Shielded Facilities.' In D. C. Stewart and H. A. Elion, Eds., *Remote Analytical Chemistry, Prog. Nuc. Energy, Ser. IX, Volume 10,* Pergamon, Oxford (1970), pp. 413-440.

22 M. T. Wilson, 'Kiloton Shield Doors at LAMPF.' (Ref. 2v above), pp. 43-51.

23 Mohammed Kaisernddin, 'Labyrinth Design in Nuclear Power Plants.' *Trans. Am. Nuc. Soc.,* **21**, 514-516 (1975).

24 W. J. Kann, S. B. Brak and J. R. White, 'Transfer Penetrations and Equipment for the HFEF Cell Complex.' (Ref. 2s above), pp. 14-19.

25 C. H. Youngquist, Argonne National Laboratory, Personal Communication.

26 C. H. Youngquist, W. C. Mohr and S. J. Vachta, 'Contamination Control in Argonne Chemistry Cave.' (Ref. 2j above), pp. 39-44.

27 R. E. Lapp and H. L. Andrews, *Nuclear Radiation Physics, Third Edition*, Prentice-Hall, Englewood Cliffs, N.J. (1963).

28 L. E. Glendenin, 'Determination of the Energy of Beta Particles and Photons by Absorption.' *Nucleonics*, **2** (1), 12-32 (1948).

29 Stanford Research Institute, 'The Industrial Uses of Radioactive Fission Products.' SRI Report 361 (1951); also published as USAEC Report AECU 1673 (1951).

30 Chapter 2, Ref. 37.

31 W. H. Steigelmann, 'Radioisotope Shielding Design Manual.' USAEC Report NYO 10721 (1963).

32 Chapter 2, Ref. 40.

33 J. H. Dunlap, P. W. Harvey and R. L. Schwing, 'Comparison of the Effectiveness of Contemporary Ophthalmic Lenses Against β Radiation.' *Health Phys.*, **32**, 555-559 (1977).

34 R. D. Evans and R. O. Evans, 'Studies of Self-Absorption in Gamma-Ray Sources.' *Rev. Mod. Phys.*, **20**, 305-326 (1948).

35 G. R. White, 'X-Ray Attenuation Coefficients from 10 keV to 100 MeV.' NBS (Chap. 2, Ref. 18) Report NBS 1003 (1952).

36 G. W. Grodstein, 'X-Ray Attenuation Coefficients from 10 keV to 100 MeV.' NBS (Chap. 2, Ref. 18) Circular 583 (1957).

37 R. T. McGinnies, 'Supplement to NBS Cicular 583.' *Ibid.* (1957).

38 C. M. Davisson, 'Interaction of β-Radiation with Matter.' In K. Siegbahn, Ed., *Alpha-, Beta- and Gamma-Spectroscopy*, Volume 1. North-Holland, Amsterdam (1965), pp. 37-78.

39 J. H. Hubbell, 'Photon Cross Sections, Attenuation Coefficients and Energy Absorption Coefficients from 10 keV to 100 GeV.' NBS (Chap. 2, Ref. 18) Report NSRDS-NBS 29 (1969).

40 E. Storm, E. Gilbert and H. Israel, 'Gamma-Ray Absorption Coefficients for Elements 1 through 100 Derived from the Theoretical Values of the National Bureau of Standards.' USAEC Report LA 2237 (1957).

41 E. Storm and H. Israel, 'Photon Cross Sections from 1 keV to 100 MeV for Elements $Z = 1$ to $Z = 100$.' *Nucl. Data Tables, Sec. A*, **7**, 565-681 (1970).

42 R. L. Walker and M Grotenhuis, 'A Summary of Shielding Constants for Concrete.' USAEC Report ANL 6443 (1961); 'Concrete Shielding Constants.' *Nucleonics*, **20** (8), 141 (1963).

43 Chapter 2, Ref. 20.

44 Chapter 2, Ref. 39.

45 J.C. Courtney, Ed., "Handbook of Radiation Shielding Data." ANS (Chap. 1, Ref. 12) Report ANS/SD-76/14 (1976).

46 N. M. Schaeffer, Ed., "Reactor Shielding for Nuclear Engineers." USAEC Report TID 2595 (1973).

47 T. Rockwell, III, Ed., *Reactor Shielding Design Manual*. USAEC Report TID 7004 (1956); also published in hardback by Van Nostrand, New York (1956).

48 H. Goldstein and J. E. Wilkins, Jr., "Calculation of the Penetration of Gamma Rays." USAEC Report NYO 3075 (1954).

49 M. A. Capo, "Gamma Ray Absorption Coefficients for Elements and Mixtures." USAEC Report APEX 628 (1961).

50 M. A. Capo, "Polynomial Approximations for Gamma-Ray Buildup Factors for a Point Isotropic Source." USAEC Report APEX 510 (1959).

51 J. P. Kuspa and N. Tsoulfanides, "Gamma-Ray Buildup Factors Including the Contribution of Bremsstrahlung." *Nucl. Sci. Eng.*, **52**, 117-123 (1973).

52 D. K. Trubey, "Gamma-Ray Buildup Factor Coefficients for Concrete and Other Materials." *Nucl. Appl. Tech.*, **9**, 439-441 (1970).

53 D. J. Rees, *Health Physics, Principles of Radiation Protection*. MIT Press, Cambridge, Mass. (1967).

54 Samuel Glasstone and A. Sesonski, *Nuclear Reactor Engineering*. Van Nostrand, New York (1967).

55 J. Moteff, "Miscellaneous Data for Shielding Calculations." USAEC Report APEX 176 (1954).

56 B. F. Maskewitz, D. K. Trubey, R. W. Roussin and F. H. Clarke, "Radiation Shielding Information Center (RSIC), a Unifying Force in the International Shielding Community," Proceedings of the Fourth International Conference on Reactor Shielding, Paris, France, October 9, 1972, pp. 215-225.

57 R. Nicks and C. Ponti, "European Shielding Information Service." *Ibid.*, pp. 226-237.

58 K. Way and K. Carver, "Directory to Neutron Cross-Section Data." *At. Data Nucl. Data*, **12**, 585-592 (1973).

59 S. F. Mughabghab, "Neutron Cross Sections, Volume I, Resonance Parameters." USAEC Report BNL 325, 3rd Ed. (1973); D. I. Garber and R. R. Kinsey, "Neutron Cross Sections, Volume II, Curves." *Ibid.*, (1976)

60 National Lead Company, "Atomic Energy Materials, 3rd Edition." (1958) N. L. Industries, Wyckoff Mills Road, Hightstown, N.J. 08520.

61 Reactor Experiments Inc., "General Catalogue 17." (1978-1979), 963 Terminal Way, San Carlos, CA 94070.

62 C. E. Kesler, Ed., *Concrete for Nuclear Reactors*. Amer. Concrete Inst., P. O. Box 19150, Redford Sta., Detroit, Mich. 48219 (1972).

63 A. N. Komarovskii, *Shielding Materials for Nuclear Reactors*, (V. M. Newton, Translator). Pergamon, Oxford (1961).

64 ANS (Chap. 1, Ref. 12), "Concrete Radiation Shields." ANS Standard ANS-11.13 1971 (1971).

65 NRC (Chap. 2, Ref. 14), "Concrete Radiation Shields." NRC Reg. Guide REG/G-3.9 (6-73) (1973).

66 F. A. R. Schmidt, "Attenuation Properties of Concrete for Shielding of Neutrons of Energy Less than 15 MeV." USAEC Report ORNL-RSIC-26 (1970).

67 H. A. Wollenberg and A. R. Smith, "A Concrete Low-Background Counting Enclosure." *Health Phys.*, **12**, 53-60 (1966).

68 Meyer Steinberg, "Concrete-Polymer Materials Program." Isot. Rad. Tech., **9**, (3) 319-321 (1972)

69 G.D. Oliver, Jr. and E. B. Moore, "The Neutron-Shielding Qualities of Water-Extended Polyesters." *Health Phys.*, **19**, 578-580 (1970).

70 C. L. Hanson and M. S. Coops, "Shielded Caves for Dual-Mode Operation." (Ref. 2t above), pp. 249-252.

71 M. S. Coops and C. L. Hanson, "The Livermore Neutron-Shielded Chemistry Cells. (1) Configuration and Philosophy." (Ref. 2o above), pp. 217-225.

72 D. H. Stoddard and H. E. Hootman, "^{252}Cf Shielding Guide." USAEC Report DP 1246 (1971).

73 D. H. Stoddard, "^{252}Cf Shielding with Water-Extended Polyester." USAEC Report DP 1339 (1973).

74   V. G. Scotti and R. L. Martin, "Construction of Hot Cells for Processing ^{252}Cf." (Ref. 2t above), pp. 253-255.

75   C. L. Hanson, M. S. Coops and E. P. Arnold, "Shielded Neutron Shipping Cask." (Ref. 2r above), pp. 162-168.

76   O. Bozyap and L. R. Day, "Attenuation of 14 MeV Neutrons in Shields of Concrete and Paraffin Wax." *Health Phys.*, **28**, 101-109 (1975).

77   T. O. Marshall, "The Attenuation of 14 Mev Neutrons in Water." *ibid.*, **19**, 571-574 (1970).

78   J. Hacke, "Dosimetry and Shielding with a 14 MeV Neutron Source." *Int. J. Appl. Radiat. Isotopes*,**18**, 33-34 (1967).

79   T. C. Gillett, R. S. Denning and J. L. Ridihalgh, "Shielding Calculation Techniques for the Design of Plutonium Processing Facilities." *Nucl. Tech.*, **31**, 244-249 (1976).

80   E. D. Arnold, "Handbook of Shielding Requirements and Radiation Characteristics of Isotopic Power Sources for Terrestrial, Marine and Space Applications." USAEC Report ORNL 3576 (1964).

81   C. P. Ross, "Cobalt-60 for Power Sources." *Isot. Rad. Tech.*, 5 (3), 185-194 (1968).

82   John Moteff, "Tenth-Value Thicknesses for Gamma-Ray Absorption." *Nucleonics*, **13** (7), 24 (1955).

83   Chapter 2, Ref. 38.

84   J. L. Balderston, J. J. Taylor and G. J. Brucker, "Nomograms for the Calculation of Gamma Shielding." USAEC Report AECD 2934 (1948).

85   L. M. Rentschler, The Polymer Corporation, Reading, Penn. Personal communication.

86   P. Monnet and J. Robert, "Miniature Radiation Resistant TV Camera." (Ref. 2w above), pp. 445-447).

87   Schott Optical Glass Inc., Duryea, Pa. 18462: (a) "Radiation Resistant Optical Glasses (Cerium Stabilized)." Brochure 7600e (1975); (b) "Radiation Shielding Glasses. Radiation Shielding Windows." (No number or date).

88   G. M. Iverson, "Wall-Periscope Viewing System." (Ref. 2t above), pp. 333-334.

89   Nuclear Pacific, Inc., 6701 Maynard Ave., South, Seattle, Wash. 98108: (a) "Penberthy Radiation Shielding Windows." (No number or date).

90   Chapter 1, Ref. 12.

91   K. R. Ferguson and R. L. Reed, "Coloration of Shielding Glasses." (Ref. 2f above), pp. 155-159.

92   E. Lell, N. J. Kriedl and J. R. Hensler, "Radiation Effects in Quartz, Silica and Glasses." In J. E. Burke, Ed., *Progress in Ceramic Science, Volume 4.* Pergamon, Oxford (1966), pp. 1-93.

93   N. J. Kreidl, "Interaction Between High-Energy Radiation and Glasses." In W. W. Kriegel, Ed., *Ceramics in Severe Environments*, Plenum, New York (1971), pp. 521-536.

94   G. S. Monk, "Coloration of Optical Glass by High-Energy Radiation" *Nucleonics*, **10** (11), 52-55 (1952).

95   B. McGrath, H. Schoenbacher and M. Van de Voorde, "Effects of Nuclear Radiation on the Optical Properties of Cerium Doped Glass." CERN Report CERN-75-16 (1975).

96   T. W. Eckels and D. P. Mingesz, "Further Data on Gamma-Induced Electrical Charge and Coloration of Shielding Glasses." (Ref 2r above), pp. 143-147.

97   V. Culler, "Gamma-Ray Induced Discharge in a Radiation Shielding Window." (Ref. 2g above), pp. 120-128.

98   J. W. Lindner, "Shielding-Glass Buildup Factors." Nucleonics, 16 (10), 77 (1958).

99   K. R. Ferguson, "Design and Construction of Shielding Windows." *Nucleonics*, **10** (11), 46-51 (1952).

100 P. W. Howe, T. C. Parsons and L. E. Miles, "The Water-Shielded Facility for Totally Enclosed Master-Slave Operations at Lawrence Radiation Laboratory." USAEC Report UCRL 9657 (1961).

101 W. B. Doe, "Zinc Bromide Solutions for Use in Shielding Windows." USAEC Report ANL 4879 (1952).

102 W. B. Lane and M. J. Nuckolls, "Argon Improves $ZnBr_2$ Shielding Windows." *Nucleonics*, **22** (2), 88–89 (1964).

103 T. E. Northrup, "A Gam-Neu Viewing Window Design." *Nucl. Struct. Eng.*, **1** (1), 108–117 (1965).

104 C. H. Youngquist and J. C. Hoe, Argonne National Laboratory. Personal communication.

105 B. D. Evans and G. H. Sigel, Jr., "Radiation Resistant Fiber Optic Materials and Waveguides." *IEEE Trans. Nucl. Scie.*, NS26 (6), 2462–2467 (1975).

106 R. A. Marbach and D. P. Mingesz, "Cell Lighting for the Hot Fuel Examination Facility—North." (Ref. 2t above), pp. 227–231.

107 R. A. Marbach, "Irradiation Tests of Hot Cell Lamps." (Ref. 2q above), pp. 222–223.

108 R. C. Goertz, "Fundamentals of General Purpose Manipulators." *Nucleonics*, **10** (11), 36–42 (1952); "Mechanical Master-Slave Manipulators." *Ibid.*, **12**, 45–46 (1954).

109 Chapter 3, Ref. 23.

110 G. W. Köhler, M. Selig and M. Salaski, "Manipulator Vehicles of the Nuclear Emergency Brigade of the Federal Republic of Germany." (Ref. 2x above), pp. 196–218.

111 Programmed and Remote (PaR) Systems Corp. 3460 Lexington Ave, North, St. Paul, Minn. 55112.

112 J. Vertut, "Virgule—A Rescue Vehicle of a New Teleoperator Generation." Second Conference on Industrial Robots, Birmingham, England, December 3–21, 1974.

113 J. Vertut, P. Marchal, G. Debrie, M. Petit, D. Francois and P. Coiffet, "The MA 23 Bilateral Servo-Manipular System." (Ref. 2x above), pp. 175–187.

114 Reference 2y above, pp. 141–173.

115 C. R. Flatau, "The Manipulator as a Means of Extending Our Dexterous Capabilities to Larger and Smaller Scales." (Ref. 2u above), pp. 47–50.

116 John McPhee, "The Developing Modular Industrial Robot." (Ref. 2s above), pp. 55–60.

117 Anon, "Blue Collar Robots." *Newsweek*, April 23, 1979, pp. 80–81.

118 J. F. Engelberger, "Robots Make Economic and Social Sense." *Atlanta Economic Review*, July–August, 1977, pp. 3–6.

119 Central Research Laboratories, Inc., Red Wing, Minn. 55066.

120 J. Vertut, J. P. Guilbaud, J. C. Germond and R. Seran, "Analytical Zoning of Manipulator Coverage." (Ref. 2u above), pp. 28–37.

121 R. G. Jeness, "Two-Piece Manipulator Boot." (Ref. 2m above), p. 187.

122 T. C. Parsons, L. E. Deckard and P. W. Howe, "An Improved Socking Technique for Master-Slaves." USAEC Report UCRL 9658 (1962).

123 C. R. Flatau, "SM-229—A New Compact Servo Master-Slave Manipulator." (Ref. 2y above), pp. 169–173. (Teleoperator Systems Corp., St. James, N. Y. 11780).

124 R. C. Goertz, R. Z. Blomgren, J. H. Grimson, G. A. Forster and W. M. Thompson, "The ANL Model 3 Electric Master-Slave Manipulator—Its Design and Use in a Cave." (Ref. 2i above) pp. 121–142.

125 Chapter 3, Ref. 59.

126 R. A. G. Welsher, "Remote Handling Equipment." In N. L. Parr, Ed., *Laboratory Handbook*. Van Nostrand Princeton, N. J. (1965), Chapter 5, pp. 2-106–2-127.

127 G. Lefort, "Critical Analysis of Solutions Used for Transfers in High-Activity Installations." *At. Energy Rev.*, **11**, 63–152 (1973).

128 C. H. Youngquist, D. J. Lind, G. A. Mack, W. C. Mohr and J. A. VanLoon, "Transport System in Argonne Chemistry Cave." (Ref. 2k above), pp. 267–279.

129 M. Wilson and L. Thorn, "Alpha-Gamma Transfer Systems." (Ref. 2i above), pp. 344–350.

130 J. Vertut, "Normalized Equipment for High-Activity Installations." *Bull. Inform. Sci. Tech. (Paris)*, **189**, 3–41 (1974).

131 O. Hladik, W. Schleicher, H. Poser, G. Schulze and R. Kirchbach, "Remote Handling Standard Laboratory Equipment." *Isotopenpraxis*, **12**(13), 133–135 (1976).

132 R. J. Smith, F. L. Sachs and W.E. Clark, Eds., "Engineering Materials List." USAEC Report TID 4100 (Suppl. 64), (July 1971); Anon, "Engineering Materials List, Cumulative Indes." USAEC Report TID 4103 (June 1968).

133 Chapter 3, Ref. 27.

134 D. E. Patterson, "Fire Protection Criteria for Nuclear Facilities." (Ref. 2r above), pp. 49–53.

135 Anon, "Rocky Flats Symposium on Safety in Plutonium Handling Facilities." USAEC Report CONF 710401 (1971).

136 A. J. Hill, Chrm., "Special Session on Fire Protection and Control in Remote Facilities." (Ref. 2r above), pp. 41–77.

137 H. A. Lee, "Final Report, Program for Fire Protection in Caves, Canyons and Hot Cells." USAEC Report ARH-ST-104 (1971).

138 A. J. Hill, "Fire Prevention and Protection in Hot Cells and Canyons." USAEC Report DP 1242 (1971).

139 A. J. Hill "Automatic Fire Extinguishing Systems for Glove Boxes and Shielded Facilities at Savannah River Laboratory." USAEC Report DP 1261 (1971).

140 Chapter 3, Ref. 25.

141 T. E. Franck and C. H. Youngquist, "Fire Protection in Chemistry Hot Cells by Use of Halon 1301." (Ref. 2o above), pp. 158–166.

142 J. P. Hughes, T. E. Franck and F. J. Schmitz, "Fixed Automatic and Manual System to Control Fires." (Ref. 2k above), pp.285–294.

143 Chapter 3, Ref. 42d.

*This shorthand notation for describing nuclear reactions places the target nucleus first, followed by a parenthetical statement. The first item within the parentheses indicates the nature of the incoming beam, the second the particles or photons that are emitted. The final notation beyond the parentheses shows the end product. Thus $^9Be(\alpha, n)^{12}C$ states that a certain fraction of the beryllium-9 nuclei struck by alpha particles of the right energy will produce neutrons and that the resulting residue will be carbon-12. The probabilities for occurrence of a particular reaction will depend on many variables. A single nucleus can have widely differing (α, n), (n, α), (γ, n), (n, γ), $(n, 2n)$, $(n,$ fission), etc., cross sections, each of which will vary with the energy of the incident beam. The predominance of a given reaction will depend on circumstances.

chapter 5

Operations

Certain daily operations are common to almost any radioactivity-handling facility. The rules under which these operations should be conducted have been codified with varying degrees of formality in many of the IAEA, NCRP, ICRP and governmental-agency publications that have been cited. The main heading for IAEA Safety Series No. 1[1] is actually "Code of Practice," and 10CFR, the portion of the U.S. Code of Federal Regulations allotted to NRC, contains many specifics relating to everyday operations.

The questions of waste disposal, transportation of radioactive materials and nuclear criticality will be taken up separately in later chapters. The emphais here will be primarily on daily routines and administration.

5.1 MONITORING

Radioactivity cannot be seen, smelled, tasted, heard or felt—its presence, character and level can only be determined by means other than the human senses. Constant and conscientious monitoring by instrumental means thus becomes the most important daily routine in any radioisotope facility since active material under nonaccident conditions is generally a problem only when its exact location is uncertain.

The instrumental and photographic techniques used for radiation monitoring are all adaptations of the same methods as used in nuclear research, although naturally not usually as elaborate in construction and exact in operation. Neither are all of the the radiation-detection options that have been applied in areas such as high-energy physics suitable for everyday monitoring, although even in this case the "bubble chamber" now appears to have shrunk to a size suitable for use as a personal dosimeter.[2] Small droplets of an organic liquid that is 10-50° superheated at room temperature are incorporated in a gel. Radiation causes measurable vaporization causing bubble formation in quantities proportional to the dosage.

The mechanisms of operation of the common monitoring devices will only be sketched here, with the emphasis being mostly on applications. Friedlander, Kennedy and Miller[3] discuss such mechanisms, as do ICRU Report 20,[4] IAEA Safety Series No. 38,[5] Attix and Roesch,[6] Morgan,[7] Shapiro,[8] and many others. Much information is also available through the commercial suppliers of monitoring instruments. Several columns of names of such firms are listed in the *Nuclear News Buyer's Guide*[9]

The various references cited above classify radioactivity detection and measurement devices in somewhat different ways. A composite view would seem to be (1) those depending on ion collection (ionization chambers, proportional and Geiger-Müller counters, solid-state detectors), (2) those depending on light emission induced by radiation (scintillation counters, thermo- and radioluminescent detectors, Čerenkor counters), and (3) those based on image forming (photographic film and cloud, bubble and spark chambers). Less common are chemical dosimeters, calorimetric methods, radiation elements, and exoelectron emission.[5] Since neutrons are uncharged, their measurement requires more indirect methods with the final determination being made by one of the abovementioned techniques after the neutron has been made to produce a detectable secondary radiation. ICRU 20, Appendix A, tabulates a large number of neutron-determination methods.

Chemical dosimeters are based on the analytical measurement of the products of a chemical reaction where the rate constants are known and there is a clear understanding of the radiation chemistry involved. Calorimetric techniques depend on conversion of the radiation to heat which is then measured directly. The various chamber techniques (cloud, bubble and spark) are basically research tools. The same is also true of Čerenkov counters and exoelectron emission. None of these will be discussed further.

Radiation causes ionization in gases and liquids, and in some solids and semiconductors furnishes enough energy to raise electrons into the conduction band. The effect in either case is to make the materials electrically conductive. This phenomenon is the basis for the great number of different ion-collection radioactivity measuring devices that have been developed. Liquid-based counters of this type are very rare, and those based on solid dielectrics or semiconductors largely restricted to specialized laboratory use. The three general types of gas-filled ion-collection instruments are however major tools for monitoring applications.

Consider a gas-filled cylinder, closed at either end and with walls whose inner surfaces are electrically conducting. The cylinder may be completely sealed and contain the gas at, above, or below atmospheric pressure, or it may be constructed so as to allow a slow movement of the gas through the chamber to be vented through a small exit aperture (gas-flow counters). The design may be such as to allow the radioactive samples to be introduced into the cylinder itself, or the latter may have a "window" at one end constructed of material thin enough to allow the radiation of interest to enter the tube from samples placed immediately below (end-window counters). Gammas of sufficient energy will of course enter though the chamber walls directly.

A conducting wire insulated from the walls is inserted down the center axis of the cylinder and a positive voltage is applied to this wire. As ion pairs are created in the gas by incoming radiations, the electrons produced will migrate to this center electrode, causing a momentary voltage pulse whose magnitude and shape when properly amplified can be used to record an ionization event and in many cases to deduce the nature and energy of the responsible radiation.

Figure 5.1, taken from Friedlander *et al.*,[3] shows the effect on the pulse height (and thus on the ease of detection) of increasing the applied voltage on the center wire of a simplified counting tube as described above. Two curves are shown, one for a particle losing several hundred times as much energy in the chamber volume as the other, as might be typical of an alpha versus a beta particle. In both cases the first application of voltage causes a rapid rise in the ion current produced by either type of radiation. In this region the induced

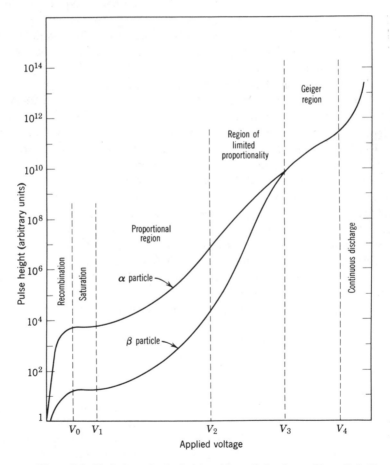

Figure 5.1   Variation of pulse height with applied voltage. (From Ref. 3.)

movements of the electrons produced by incoming radiation are essentially so slow as to allow a varying fraction of recombination with positive ions before the center wire is reached. Both curves level off between V_0 and V_1, the saturation current region. Instruments of the ionization-chamber type operate over this plateau. Pulses due to individual beta particles are too small to be detected separately with practical electronics, although the alpha pulses can be separately distinguished with a sensitive pulse amplifier. In either case, the integrated effect of many ionization events will produce enough ion current to be measured and correlated with exposure.

As the voltage continues to be increased above V_1 the electrons produced by an incoming particle or photon are themselves accelerated towards the center wire with enough energy to cause ionizations of their own as they pass through the gas, the net result being an "avalanche" of electrons striking the wire and thus increasing the resulting pulse height by orders of magnitude. In this "multiplication" range (V_1 to V_2 of Figure 5.1) the *ratio* of pulse heights caused by different types of particles is independent of the applied voltage, that is, the pulse height remains proportional to the amount of energy lost in the chamber by the primary ionizing particle. This then is the voltage range utilized in proportional-counter design. With these instruments betas can be distinguished in a gamma background and alphas counted in the presence of betas.

As voltages are increased above V_2 the pulse heights continue to increase, but proportionality is gradually lost (the entire electrode rather than just a portion becomes involved in the discharge), and at V_3 the pulse height becomes independent of the original ionization, making it no longer possible to distinguish between types of initiating radiation. Pulse heights in this Geiger-Müller (G-M) region are so high as to require very little additional amplification. Above V_4 the counter goes into continuous discharge and practical application is no longer possible.

The health physicist in even a medium-sized organization is likely to encounter monitoring problems involving various types of radiation from weak x-rays to fast neutrons. Each of the three types of gas-filled detectors will be more applicable in certain situations than the other two, so a variety of monitoring instruments based on all three (along with the scintillation monitors discussed below) is almost a necessity. Ionization chambers use lower voltages, are more reliable in extremes of temperature or humidity, can be rugged and simple in design and can be readily modified to read an absorbed dose directly without the necessity of converting the readings. The instrument is however relatively insensitive unless the chamber volume is large or the gas pressure high. Proportional counters are very much more sensitive and are particularly useful when both alpha and beta emitters are being handled since the contribution from each can be readily determined. This is easily seen in Figure 5.2, also taken from the text by Friedlander and his associates. In the example shown, only alphas will be seen over roughly the 1450–1850V range whereas alphas plus betas are measured between 2100 and 2700 V. Insulator problems may be difficult, especially under conditions of high humidity. The instrument can be

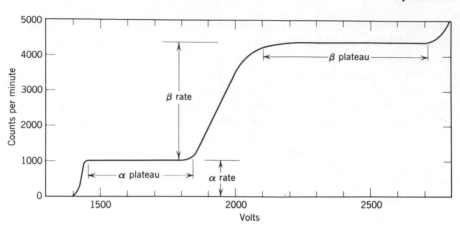

Figure 5.2 Counting rate in a proportional counter as a function of applied voltage. (From Ref. 3.)

used in either the counting or current-measurement mode and can be adapted as a spectrometer for measuring alpha energies if suitably designed.

G-M counters are very sensitive, but will not distinguish between different types of radiation. Earlier G-M tubes had limited working lifetimes, but models are now available ("halogen quenched") that have largely overcome that difficulty and are operable at somwhat lower voltages. The dead time (the time needed for an electron avalanche to be initiated, to travel along the wire, and to be quenched leaving the counter ready for the next pulse) is longer than in the other instrument types, so the G-M counter is usually used when the counting rate is less than a few thousand events per minute. In more intense fields the counter may "block," giving no reading at all. The associated electronics may be affected by extremes of temperature and humidity. "Even so, the electronics of G-M counters are subject to fewer difficulties than those of other systems because the high signal levels generated by G-M tubes permit the use of simple, rugged circuitry."[4]

The most senior of the measuring techniques based on induced light emission is the zinc sulfide screen for detecting alphas, a method almost as old as the discovery of radioactivity. An alpha particle striking the screen generates a visible flash of light and by counting the number of flashes per unit of time, early investigators were able to calibrate the intensity of their sources. It was later found that betas and gammas would similarly produce flashes in anthracene and other organic crystals. This technique then led to the development of liquid-solution detectors such as p-terphenyl or stilbene dissolved in xylene or toluene, and there are many such "scintillation cocktails" available commercially. Compounds containing soft beta emitters such as ^{14}C and ^{3}H are often dissolved directly in the solution for assay. If the cocktail is loaded with boron or cadmium salts it becomes a good slow-neutron detector because of the secondaries produced by the respective (n, α) and (n, γ) reactions. The workhorse of the scintillators for gamma counting is the thallium-doped sodium iodide crystal. The pulse height seen by the electronics varies with the energy of the gamma generating the light flash, allowing the NaI(T1) counter to be adapted as a gamma-ray spectrometer.

The emitted light in all of the scintillators is directed onto the first photosensitive electrode of a photomultiplier tube, which device increases the output pulse by orders of magnitude. There have been constant improvements in such tubes and in the other supplemental electronic equipment over a number of years.

Some materials, such as the fluorides of lithium and calcium, store energy when exposed to radiation. This energy can at a later time be released in the form of visible light by heating. This is thermoluminescence. Other materials, such as silver-activated phosphate glass, not ordinarily luminescent under ultraviolet light, become activated after exposure to x-ray or gamma radiation. The intensity of the induced ultraviolet flourescence is again related to the amount of exposure, so such materials can be used as dosimeters. This last phenomenon is radiophotoluminescence.

As Becquerel rather unintentionally demonstrated in 1895 when he discovered natural radioactivity, radiation will cause darkening of photographic plates or film. This effect was widely used in early research as an investigative probe and of course is still a major tool for the medical and dental professions. The technique fell into relative laboratory disuse for a time after the advent of the more convenient electronic measuring devices, but has made somewhat of a comeback with applications in radioautography and neutron and gamma radiography, as well as in dosimetry. The emulsions used are higher in silver halide content and smaller in grain size than ordinary film. As the grain size decreases, the emulsions become less sensitive to any but the most densely ionizing particles. Emulsions differing in grain size are commercially available that allow distinguishing between different types of particles.

Monitoring instruments can also be classified in terms of application: (1) equipment in fixed locations (gamma-level sensors around or in hot cells, doorway monitors, hand-and-foot counters), (2) devices worn by individual workers (film badges, dosimeters, "chirpers") and (3) portable instruments for on-the-spot measurement of radiation fields or for surveying for contamination. In a large organization these will be backed up by a laboratory or laboratories furnished with nonportable counting instruments such as a gamma-ray spectrometer for identifying contamination components of uncertain origin. There may also be a group established for examining urine and fecal samples from the individuals handling the radioactivity, and perhaps a whole-body counter. Environmental monitoring[10,11] around a reactor or a large plant may depend on a mixture of fixed field instruments and periodic surverys with portable equipment; with soil, water and vegetation samples taken at scheduled intervals and brought back to the laboratory for analysis. Many services of these last types are available commercially for purchase by organizations whose involvement with radioactivity is not large enough to justify an elaborate in-house back-up system. (As an example, ICRU 20 estimates that at least 20 personnel film badges per developing session are needed before it becomes economical to invest in the needed facilities.)

Fixed gamma sensors are usually adjusted so that they actuate audible and visible alarms when the activity level exceeds a predetermined value. They may also be connected to a central console where the existing level at any time can be read from a dial. Doorway monitors position several detectors around an exit opening, but are generally more for security than health physics purposes as a means of detecting theft of radioactive material. Criticality alarms[12] may have a rate-of-rise feature incorporated since they are meant to sense a very sudden event. Those that depend entirely on gamma detection may produce a false alarm if a highly active source is inadvertently moved too close, but types are now available where fast neutrons[13] or both neutrons and gammas must be seen before the alarm is actuated. Exhaust stack monitors indicate any failure in the off-gas treatment system and are particularly important to reactor fuel reproccessing plants during the preliminary dissolving step when the bulk of the fission gases in the fuel is released. An entire session of the twelfth AEC Air Cleaning Conference[14] was devoted to stack monitors, and individual papers abound in the proceedings of all of the meetings of the series.

Contamination tracking incidents do not happen every day and almost invariably pose no particular threat to health, but the activity must be removed, a disruptive operation if the tracking has been extensive. Hand-and-foot counters, preferably in depth, are by far the best defense, making these instruments probably the most important of the fixed-position monitors for everyday operations. They should be placed in critical locations throughout the working area, and, most particularly, at all exits. Their use by everyone leaving the area should be mandatory with some sort of disciplinary action taken against those who chronically ignore the machines.

Figure 5.3 pictures a commercially available hand-and-foot counter. Gas-filled detector chambers do not necessarily have to be cylindrical in shape, and in counters of this type (and in many portable survey instruments) are made as very shallow boxes with sufficient window area to completely cover an ordinary hand or shoe. The window is a foil (frequently double-aluminized Mylar) thin enough to permit the passage of alphas or betas. This film is protected by a perforated metal grating. In the model shown in the figure the user stands on the detectors in the base, inserts his hands into the openings in the top section, and presses downward. This action triggers a count for a preset length of time. Lights on the face of the instrument then indicate either that there is no contamination, or, if there is, whether it is on a hand or foot, whether left or right and whether alpha or beta. A telephone or some other means of alerting the monitoring group should be immediately at hand so that if there is contamination its unwilling host will not be forced to move and thus spread the problem further before getting assistance.

Hand-and-foot counters are usually of the gas-flow proportional type so that alphas can be detected. The commercial versions are not inexpensive be-

Figure 5.3 A commercially available hand-and-foot counter. (Courtesy Eberline Instrument Corp., Ref. 15.)

cause of their versatility. Simpler devices[16] will serve in some specialized situations. Other instruments have been described.[17,18] A hand-and-foot counter receives rather rough treatment so its functioning must be watched. The chief maintenance problem is probably small leaks in the Mylar-film windows of the foot units since these are continuously exposed to dirt and other debris from the shoes of the users. A hole in the upstream probe means that gas flow to the downstream detector is broken, making the unit inoperable.

Pocket dosimeters are fountain-pen-size devices furnished with a clip so that they can be pinned to a laboratory coat or carried in a pocket. The most common types are small ion chambers, in the simplest version of which a known charge is put on the central electrode before the monitor is assigned to a wearer. Depletion of this charge is measured after use to determine the integrated exposure. A more elaborate version incorporates an electroscope by attaching a metallic or metallized quartz fiber to the electrode. The initial charging causes this fiber to be repelled from the electrode, while ionization in the chamber during exposure allows it to move back. The fiber tip travels across a graduated scale that can be observed through a built-in lens system. The scale covered is usually over the 0-200-mR range, but units can be obtained to show integrated dose up to 5 R. Such dosimeters thus provide an immediate summation of all the exposure received since it was charged. Quartz-fiber dosimeters are however easily damaged and should always be backed up by a second monitor, usually a film badge as discussed below. The quartz-fiber dosimeters are basically to provide immediate information to the wearer. The readings from the film badge provide the data going into the exposure records.

Physically slightly larger pocket dosimeters are also available in which an ionization in the unit is converted to an audible sound. These "chirpers" or "beepers" are very useful around a hot cell since the wearer is immediately warned if he strays into a high radiation field.

Thermoluminescent and radiophotoluminescent pocket dosimeters are available and have the advantage of being relatively insensitive to humidity, electric and magnetic fields, and the presence of air-borne chemicals. There has been limited experience to date in utilizing these solid-state detectors in large-scale monitoring programs, but there is every reason to believe that their use will grow.[4] These types of dosimeters should have special appeal to smaller groups of radioactivity users since darkroom facilities are not required for their development.

The second major monitor worn by personnel is the film badge. The film packet is held in a metal or plastic frame a few inches on a side and $1/4$ in. thick. The frame is again furnished with a clip so that it can be attached to clothing or carried in a pocket. The film is enclosed in a light-tight jacket having a small window so that a portion of the film is exposed directly as it is held in the badge frame. Other parts of the film are shielded on both sides by thin foils of different materials. Some idea can then be gained concerning the types and energies of the exposure radiation by comparing the darkening in the shielded and unshielded areas after film development and relating this information to the ratios of the shield densities. IAEA Safety Series No. 38 shows a six-compartment badge—one section open and others shielded with plastic, lead, cadmium and copper in two different thicknesses. Film-badge development and interpretation is a specialized activity and even larger organizations often contract for the service with an outside laboratory.

The film badge, as with the pocket dosimeter, is usually worn on the trunk of the body. Film badges can also be exposed in small frames attached to wrist bands or finger rings. (Types are now available in which the film is enclosed within the ring itself.) The hands and fingers are likely to be the most heavily exposed parts of the body in laboratory work and such specialized film badges are among the few ways available for evaluation of such exposure.

Film gadges and dosimeters are usually kept in a rack at the entrance to a controlled area when not in use. Each regular employee is assigned a pocket dosimeter and a badge which are picked up when reporting for work and returned to the rack when leaving. The monitors are all individually numbered and assigned, and the badge may carry the employee's name. The monitoring group periodically collects the exposed monitors and replaces them with fresh units. The collected devices are then read and the data obtained placed in the individual's exposure record, with specific investigative actions mandated if an unusually high reading is found. The collection and reading cycle will vary depending on the nature of the program, but is often on a once-a-week basis. Visitors to the controlled area should be given numbered pocket dosimeters and asked to complete cards giving their names and affiliations. The dosimeters are returned upon departure and the readings from them marked on the card, after which it is filed. This procedure is of course a legal as well as a health physics precaution. One or two dosimeters may suffice for a group of visitors if they stay together.

Portable survey instruments are typically needed either to measure on-the-spot radiation fields or for detecting surface contamination. A great variety of excellent devices is now available commercially for very many different specialized applications. Popular forms for more general work may be all in one package, such as the hand-held "gun-type" monitors, or may have the detector probe connected to the electronics cabinet by a cable so that the probe can be handled separately for awkward jobs such as surveying shoe soles or getting into corners. The electronics unit may be light enough to be hand carried, or may be placed on a laboratory cart, particularly if the probe requires a counting gas supply. A useful arrangement of this type allows a single electronics plus gas-supply unit to be connected to and used with several different types of probes.

Most of the survey instruments are one of the three gas-filled ion-collection types, each with its own advantages and limitations. The G-M counter is highly sensitive but easily overloaded, so it is usually used first to determine whether radioactivity is or is not present. A switch might then be made to an ion-chamber type for completing the mapping of a gamma-exposure field or for searching for radiation leaks in a newly built shielding wall if the radiation level found could cause overloading of the G-M. A proportional counter could be used if possible contamination were the problem and alphas could be present, surveying first in the alpha mode, then for total alpha plus beta to complete the picture. This last technique is particularly valuable in using floor monitors. (These are proportional counters mounted with the detection surface face down in a device resembling an old-fashioned carpet sweeper and having a readout dial on the handle. Such monitors are almost a necessity in a building where large areas of floor space must be surveyed either routinely or after a contamination tracking incident.) The voltage can be raised if the original sweep indicates that alphas are absent and the survey repeated in search of betas.

Air-borne activity in a controlled area is a matter of serious concern, particularly during operations involving the transfer of materials in and out of enclosures. Stationary monitors can be installed in areas where large amounts of active materials are handled on a routine basis such as in a glovebox facility for fabricating plutonium fuel elements. Designs differ, but usually air is drawn at a preset rate through a filter medium whose surface is monitored by a proportional or scintillation detector. Adjustments have to be made[19] for the fact that such filters will accumulate active daughters of the natural radon and thoron decay chains. Any surge of activity above this background sets off an alarm. A somewhat similar but simpler approach is used in the laboratory. Small pumping units, wheeled or with handles so that they can be readily moved from place to place, are used to draw measured quantities of air through a small filter which is removed and replaced at specified intervals. The activity on the filter is then analyzed in a bench-type counter. An equally valuable and straightforward technique is often used for detecting contamination on surfaces where direct instrument surveying is difficult because of the existing background or where the amount of loose contamination is below the sensitivity detection level of the portable instruments. A piece of paper tissue or a wad of cotton is "swiped" over the surface and the swipe then counted in a laboratory away from the background.

The portable survey instruments must be constantly kept in calibration, particularly if there is a need to obtain quantitative estimates of exposure. Such calibrations can be reasonably straightforward if only a single nuclide is being handled, but can become involved if varying radiation fields of mixed energies are routinely encountered. Sealed isotopic sources of known activity are usually carried along with the survey devices and occasional checks of instrumental response made against these sources to be certain that the equipment is functioning normally. The matter of instrument calibration is taken up at some length in ICRU 20, IAEA 38, Attix and Roesch and other references previously cited in this section.

Figure 5.4 pictures some of the monitoring devices that have been discussed in the previous pages. The detector area for the G-M survey meter shown in Figure 5.4 (*a*) covers most of the underside of the instrument, making it very useful for contamination monitoring of flat surfaces. The meter in Figure 5.4 (*b*) has a detachable probe which can be of either the G-M or scintillator type. Figure 5.4 (*c*), "Cutie-Pie," is of the gun type. Figure 5.4 (*d*) shows a direct-reading quartz-fiber pocket dosimeter, and Figure 5.4 (*e*) a chirper. Figure 5.4 (*f*) is a photograph of a portable low-volume air sampler.

A health physics group will have many nonroutine assignments such as review of the radiological safety aspects of new projects and buildings, advising on appropriate instrumentation for new situations, investigation of incidents, training of radiation workers and its own research and development programs. There are also very many routine procedures to be carried out on predetermined schedules. Maintenance and calibration of the various protective instruments is a perrenial and extremely important function. The system for

(a)

(b)

Figure 5.4 Radioactivity monitoring instruments: (a)-(c) Survey meters. (d) Pocket dosimeter. (e) Chirper. (f) Portable air sampler. (Courtesy Victoreen, Inc., Ref. 20.)

167

(c)

(d)

Figure 5.4   (Continued)

assignment, collection and reading of personal dosimeters and film badges has already been mentioned, as has the collection and measuring of air samples. Floors should be surveyed on a regular schedule, as should be other laboratory areas, particularly the sinks, waste containers and the hood fronts. Surveillance of proper signs and labels for locations of active materials must be maintained. The mops and other equipment used by the janitorial staff in controlled areas should be checked as an indication of contamination spots that might have been missed. The controlled-area waste-disposal system must be continuously monitored and made safe. Laboratory clothing worn by the radiation workers should not be placed in lockers after use, but left on hooks in the open on the hot side of the change room so that it can be monitored each day, particularly before being sent to the laundry. All items leaving the Zone-3 and Zone-4 areas must be thoroughly surveyed before being allowed out. Any suggestion of possible overexposure or contamination must be quickly investigated. And there is always a near infinite number of records to be kept current.

Possibly the most important single activity of the monitoring group is still that of being on hand during transfers. Radioactivity housed in a properly

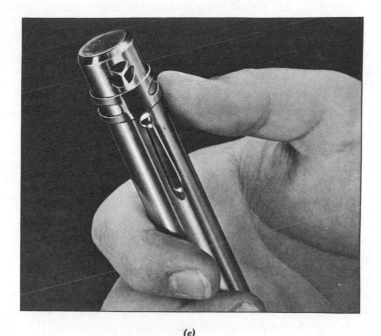

(e)

(f)

Figure 5.4 (Continued)

designed box, cell, shipping container or storage unit is normally no problem; the chance for accident or contamination arises when the enclosure must be breached to move materials in or out, or even in simply moving items from one hood to another. Careful monitoring during these operations can avoid many later problems. The radiation worker bears an equal responsibility in these cases since the monitoring group cannot do its job if it is not aware that a transfer is being made. Monitors should also be on hand at any time when laboratory furniture is being moved, plumbing and repairs being made, or obsolete facilities or equipment being dismantled. IAEA Safety Series Nos. 1 and 38 both itemize routine monitoring activities as does NCRP No. 10,[21] Shapiro,[8] Rees,[22] Moe *et al.*,[23] the Radiological Health Handbook[24] and many of the other references that have been cited.

ICRU 20 has been mentioned previously as a summary source of the specialized techniques needed for monitoring for slow and fast neutrons. Another large area of specialization is in surveying for low-energy x-rays and for the very soft betas emitted by such nuclides as ^{3}H, ^{14}C, ^{99}Tc, ^{147}Pm and ^{249}Bk. The references to tritium[24] and carbon-14[25] given in Chapter 3 detail monitoring techniques for these last two nuclides. Brown *et al.*[26] have described a field air sampler for tritium, and an Oak Ridge bibliography[27] on the isotope is available that lists other pertinent references. Ello[28] has reported on a simple modification of a commercially available proportional-type survey instrument to make it suitable for detection of very soft betas.

5.2 DECONTAMINATION

Decontamination is an unavoidable fact of life when handling radioactivity, even under the Concentrate and Confine philosophy since enclosure interiors will eventually get to the point where some cleanup in necessary if materials are to be safely moved in and out. The arts of decontamination are moreover increasing in importance as first-generation nuclear facilities become obsolescent and candidates for decommissioning, a step mandating the elimination of as much residual activity as possible. "Decommissioning and Decontamination" has been the topic of various conferences[29,30] and of an increasing number of reports.[31,32] These of course deal primarily with fuel-cycle facilities and are somewhat beyond the scope of this book, but many of the techniques used and the problems met are closely related to everyday laboratory and hot cell operations.

Decontamination should be thought of from two points of view. If an item, room or facility is to be released for subsequent unrestricted use a very stringent set of rules must apply, as is also the case for Zone-1 areas in nuclear installations or when cross contamination from one experiment can ruin another. On the other hand, if the item or space is in a controlled area, part of a continuing program, and will be put back into use in much the same manner as previously after being cleaned, decontamination need be only to the point of ensuring safety and of simplifying operations such as transfers and equipment replacement and repairs. Different decontamination operations thus have different aims, a fact that probably has much to do with the lack of a universally accepted international standard for "how clean is clean?"

Most organizations thus establish their own internal standards for allowable levels of radioactivity on surfaces. Schmidt[33] has reviewed surface-contamination limits established by different groups based on examination of available guides and answers to a questionnaire. IAEA SS No. 1 summarizes some European standards. France, Poland, the United Kingdom and Switzerland are in agreement on 10^{-4} μCi/cm^2 as the limit for the high-toxicity group of alphas, usually with a factor-of-10 less-stringent limits for other alphas and for beta-gamma emitters. This value would seem to be derived from the early recommendation of Dunster[34] of a limit of 22,000 (dis/min)/100cm^2. The Russian standard for highly toxic alpha emitters on working surfaces is only 1000 particles/100 cm^2 but 200,000 for beta emitters.

Recommended standards for materials to be released for subsequent unrestricted use were set forth in this country in NBS Handbook 92[35] in 1964 with the numbers stated in terms of the monitoring instrument used. The only directly official U.S. standard published to date appears to be NRC Regulatory Guide 1.86[36] dealing with acceptable limits for surfaces in decommissioned reactors. These values are given in Table 5.1. It will be seen that they are more conservative than the European limits, but again the unrestricted use versus continuing program difference in intent must be taken into account. ("Removable" in Table 5.1 indicates the quantity obtained by a single swipe of the surface. The maximum values would be acceptable only for a limited number of small "hot spots," each of less than 100 cm^2 in area.)

In practice, the method of reporting and recording surface contamination often depends on the face area of the monitor being used. Many such instruments are also equipped with thin aluminum shutters to shield out soft betas and alphas, thus permitting a double reading of the same area. A beta-gamma survey then might be reported as, "2000 c/m hard, 8500 c/m hard and soft per 61 cm^2"

While the limits in Table 5.1 currently apply only to reactors, the NRC has prepared a draft[37] based on the same limits to cover all facilities and equipment

Table 5.1 NRC Standard for Decontaminated Reactors

	(dis/min)/100 cm^2		
Nuclide	Average	Maximum	Removable
U-nat, ^{235}U, ^{238}U, and decay products	5000 α	15,000 α	1000 α
Transuranics, ^{226}Ra, ^{228}Ra, ^{230}Th, ^{228}Th, ^{231}Pa, ^{227}Ac, ^{125}I, ^{129}I	100	300	20
Th-nat, ^{232}Th, ^{90}Sr, ^{223}Ra, ^{224}Ra, ^{232}U, ^{126}I, ^{131}I, ^{133}I	1000	3000	200
Beta-gamma emitters except ^{90}Sr and others noted above	5000 $\beta-\gamma$	15,000 $\beta-\gamma$	1000 $\beta-\gamma$

Source: Condensed from USNRC Regulatory Guide 1.86 (Ref. 36).

scheduled for release for unrestricted use. NRC inspectors operate on the basis of this draft. For ongoing operations a limit of 100dpm/100 cm^2 is required for clean areas. Limits in restricted access zones are applied with considerable flexibility, depending on the situation. Both the NRC (10CFR20) and DOT (49CFR173) restrictions for surface contamination on shipping containers are based on the Dunster 22,000 dpm/100 cm^2 limit.[38]

The fundamental processes involved in surface contamination and decontamination are not fully understood. An older paper by Stevenson[39] presents information that is still valid. He believes the single most important absorptive process that occurs is that of ion exchange, with physical adsorption normally being of lesser importance. Partly because of their inherent nature, and partly because of surface oxidation, all organic and vitreous surfaces contain acidic groups. These ionize above pH 2 and act as low-capacity ion exchangers. Anions will be repelled unless a specific chemical reaction occurs, such as iodine reacting with a double bond. After exchange occurs, mass-action replacement with high concentrations of nonactive ions will have some effect, particularly if the replacement ion is polyvalent, such as Al^3 or Fe^3 . The efficiency of the mass-action approach is however low as compared to the use of complexing agents. EDTA (ethylene-diaminetetraacetic acid) is particularly suitable, but the pH of the solution must be controlled. EDTA seems to be most effective in a citrate-buffered solution of fairly low pH and with a wetting agent added. The citrate appears to act synergistically in improving the action of the EDTA.

Polyvalent cations may penetrate past the surface ion-exchange layer if the contamination occurs under highly acidic or ionic conditions. Decontamination becomes very difficult if such diffusion occurs, and the situation is even worse if the contaminants can actually dissolve in the material. Many curing agents and plasticizers in polymeric compounds form oil-soluble complexes with heavy metals under acid conditions. The material generally can be cleaned only by removal of the surface layers if subsurface contamination of this type takes place, preferably by attritive methods since attempts to dissolve away the surface will probably only result in driving the activity further into the material.

The exchange action generally occurs with the external oxide layer in the case of metal surfaces, with copper being a particular offender. The amount of exchange for bare metal surfaces will be determined by the relative positions in the electrochemical series of the contaminating cation and the reacting metal. Most metals rapidly become covered with a thin film or grease in industrial areas. This coating gives some protection, but itself is readily contaminated. Detergents or degreasing solvents are useful in such situations.

Stevenson believes that many decontamination methods are more drastic than need be, and that every effort should be made to use the mildest possible conditions. He suggests that a natural rubber latex be sprayed on a surface, then stripped away to remove much contamination. Adhesive tape can be applied and removed to give a similar effect. This technique is even more effective if a solution of a detergent and a complexing agent is allowed to dry on the surface before the latex or tape is applied.

Many decontamination operations follow a similar pattern. A large fraction of the activity is initially rather easily removed, following which a plateau occurs in which each new treatment takes off a small and relatively constant amount. This eventually tapers off to the point where very little new material is removed per round, although measurable activity still remains. Presumably this last tenacious fraction has gone below the surface, perhaps aided by the recoil mechanism that has been referred to previously. It is therefore very worthwhile to decontaminate as early as possible. There is considerable evidence showing that activity that has been on a surface for but a short time is much easier to remove than when the contamination is of long standing.

Since the basic mechanisms of contamination are only partially understood, research on the phenomena has largely been empirical. Several bibliographies are available.[40-42] Oak Ridge conducted an extensive experimental program in the 1950s and 1960s[43] and the Kernforschungzentrum at Karlsruhe has a very active group currently working on decontamination problems and has issued a number of reports.[44]

The *Radiological Health Handbook*[24] devotes a number of pages to detailed applicable procedures, dividing the section into (a) personnel decontamination and (b) area and material decontamination. In each case the recommendations begin with milder methods then become increasingly more drastic if the contamination persists. For skin and hands the ordering is: soap and water; lava soap, soft brush and water; a detergent solution; a 50-50 mixture of detergent and corn meal; a 5% water solution of 30% Tide, 65% Calgon, and 5% Carbose (carboxy methyl cellulose); a mixture of 8% Carbose, 3% Tide and 1% Versene in water; titanium dioxide paste; and finally a 50-50 mixture of saturated potassium permanganate and 0.2N sulfuric acid, to be followed by a 5% solution of sodium acid sulfite to remove the permanganate stain. Precautions to be taken with these later and harsher mixtures are outlined in the reference. Contaminated hair should be thoroughly washed with soap and water. If this is not effective the hair may have to be shaved off and the various skin treatments then applied. Contamination in eyes, ears, nose, mouth and open wounds should first be treated by flushing with copious quantities of water. Contamination on hands and feet may sometimes be removed by stimulating heavy sweating. Rubber or plastic gloves or booties are taped around the member which is then held close to a source of heat.

The approach used for areas and equipment is necessarily highly dependent on the nature and extent of the surface to be treated. It must be kept in mind that methods that involve large volumes of liquid may also generate large volumes of contaminated waste that will require disposal, and that if any equipment is to be re-used, the decontamination cannot be so strenuous as to leave the item in inoperable condition. No effort will be made to summarize the extensive list of methods listed in the *Handbook*, but general comments will be made based on those recommendations and other sources such as IAEA Safety Series No. 38,[5] Watson *et al.*,[45a] Schneider *et al.*,[46] Ferguson,[45b] and Wells.[45c]

The first step in decontaminating boxes, cells and hoods is usually to take out all the removable items for separate treatment or disposal as active waste. In many cases this step is associated with a thorough vacuuming or sweeping of the enclosure, particularly if powdered materials have been handled. Any protective strippable material present (see below) is also removed. These actions are frequently sufficient to reduce the background to the point where a realistic mapping of the location of the remaining radioactivity can be made.

Decontamination methods beyond these preliminary stages can be roughly classified as chemical or attritive. The latter employ various devices to physically remove surface material to a depth where no more radioactivity is seen, and will almost certainly have to be applied at some stage if the contamination is of long standing on bare concrete or a similar porous surface. Sanding, filing, chipping, paint removal, sandblasting, grinding, and even shaped explosive charges when decommissioning large facilities may have to be used. Much care must be taken because of the hazard of contaminated dust and the possibility of even further spread of the activity.

Some shielded cells are designed with hot drains so that high-pressure water jets or steam can be used as one of the first decontamination steps. Cleaning begins with the ceiling, then works downward on the walls. In-cell filters and electrical equipment must be protected. Hoods and boxes can be similarly swabbed out with sponges, using detergent and water. These steps again are somewhat preliminary since they are rarely sufficient by themselves to eliminate all of the activity. Chemical treatments are then used, their character depending on the nature of the surface and the contamination.

IAEA 38 recommends either an unbuilt detergent (no admixture with inorganic salts) or a recipe of 5%-10% Na_2CO_3 or built detergent plus 1%-2% EDTA and 0.5%-1% citric acid for simple decontamination of glass, stainless steel, copper, aluminum, lead, and rubber. If it is desired to remove some surface as well, chromic acid is used on glass, a 1%-H_2SO_4-25%-inhibited H_3PO_4 mixture for stainless steel, sodium hydroxide or citric acid for copper and aluminum, aqua regia for lead and acetone for rubber.

Walker and his associates surveyed American experience, primarily from the point of view of cleaning up hot cells. They list specific mixtures for decontaminating painted surfaces and the common metals. Many of their recipes contain commercial cleansers, a number being manufactured by Turco Products, Inc.[47]

The *Radiological Handbook's* long list is not generally as specific regarding solution compositions, but contains considerable detail on methods of application. Water, steam, detergents, complexing agents, organic solvents, inorganic acids and acid-salt mixtures as decontamination agents are all reviewed. Strong caustic solutions can be used to soften paint on horizontal surfaces. Tri-sodium phosphate in a hot 10% solution can be used for the same purpose in a somewhat milder treatment.

Decontamination is a tedious and labor-intensive operation. Unless an item can be reused after only a cursory cleaning, it is frequently more economical simply to replace it. The high cost in time and dollars of decontamination also makes it very worthwhile to preplan for the day when the operation must take place. There are a number of steps that can be taken and here the old cliché regarding an ounce of prevention is highly valid. As one general principle, the amount of surface exposed to heavy contamination should be minimized as far as possible. A separate small enclosure should be placed around the cell compartment of a spectrophotometer, for example, and ideally every analytical instrument should have only the sensor portion inside a hood, cell or glovebox while all the other parts of the equipment remain outside. Heating operations that could produce contaminated fumes should be done in closed systems and particularly dusty operations performed within a separate box within the main enclosure if feasible. The containment-box approach to handling alpha emitters within a hot cell as described earlier is an excellent application of the last two principles.

A second general rule is to furnish an expendable surrogate surface to collect the contamination in order to protect the underlying more permanent parts of the facility. The present ready availability of various types of plastic sheeting helps considerably, as can readily strippable paints and coatings. Trays and other secondary containers should be used on the working floors of hood, box and storage units in the laboratory. Bench tops outside the enclosures can be covered with polyethylene sheet taped into place, as can the more vulnerable floor areas in front of hoods, boxes and storage safes. Plate-glass placed over resilient and absorbent paper sheeting, with all cracks sealed with industrial tape, makes an easily decontaminable and generally satisfactory working surface, although care must be taken in the placement of heavy or heat-producing equipment. The metal portions of items that must go into a hot enclosure can be given an epoxide-based coating or covered with an easily removed paint. The larger equipment more characteristic of hot cells can be given similar treatment, or even, "cocooned" with a strippable coating based on vinyl chloride–vinyl acetate copolymers.[45c] Liberal use of polyethylene sheeting as protective covering within a cell can also help as long as it does not interfere with operations, although its working lifetime is likely to be rather short in high activity fields. The moving parts of equipment can in some cases be given a heavy coat of protective grease. The intent throughout is to substitute a relatively easily removable surface for one that is more difficult to decontaminate.

The interior walls of many large hot cells are either lined with stainless-steel plate or are of heavily coated concrete. A strippable paint is indicated if steel is used, one soluble in water-steam[48] if the proper disposal facilities for the waste are available. Concrete walls should first be covered with a highly impermeable paint giving a hard and smooth surface, then given a strippable outer coating. Smaller shielded facilities of lead brick will have many fine cracks on the inner surface where contamination can lodge. Painted steel inner liners are sometimes used to solve this particular problem, or a heavy paint can be applied to the lead directly. Many facilities have established separate large gloveboxes or dedicated shielded cells specifically for decontamination of portable items. Sonic cleaning with the object submerged in a detergent or other decontamination solvent can sometimes be effective for small equipment having such irregular surfaces as to make them almost impossible to clean by other techniques.

The importance of specifying decontaminable surfaces in designing a laboratory or hot cell has been emphasized in previous chapters. Wells[45c] and West and Watson[49] have both conducted studies of the resistance to radiation damage of various polymeric coatings and of the relative ease of decontaminating those surfaces both before and after the exposure. A large number of compounds were tested in both investigations, some commercially available and others prepared by differing curing techniques. Epoxy resins, phenolic resins and inorganic zinc coatings had the best radiation resistance, whereas the vinyls and glossy epoxies were superior in terms of decontaminability.

5.3 PERSONNEL PROTECTION

The subject of decontamination leads naturally to discussion of protective clothing, broadly interpreted here to include respirators and similar devices for promoting individual safety.

Safety glasses, safety shoes, face shields, fire blankets, eye-wash devices, etc., are normally present in most laboratories, and the use of such items is assumed. Over and above such equipment, ordinary laboratory coats, aprons or smocks offer adequate protection for the occasional user of low levels of radioactivity. Gloves of some sort should be worn while actually handling active material, and hands, shoes and personal clothing not protected by the outer garment monitored after the job is completed. The coats themselves must be similarly checked before being laundered.

More elaborate systems become necessary as the frequency and level of the radioactivity handling increases or when the high-toxicity nuclides are being used. This is usually accomplished by furnishing shirts, trousers, coveralls and safety shoes that are worn only during the working period and left in a change room for monitoring after the individual has changed back into his own personal clothing. Some organizations go so far as to also furnish socks and underwear and IAEA Safety Series No. 22[50] has several pictures of a rather self-conscious appearing mannequin wearing highly utilitarian undergarments of this type. Cloth, plastic or paper head covering may be indicated for some operations, or hard hats for others.

Establishment of such a clothing system has several implications. Change rooms must be furnished for male and female employees, most reasonably associated with shower and toilet facilities. Each change room should be divided into a hot side and a cold side, preferably with the showers in between. The work clothing should be removed on the hot side and left on hooks to facilitate monitoring, and shoes placed in an appropriate storage rack. A bin for disposable shoe covers should be furnished if these are used. The clean side should have an individual locker for each employee's street clothing and a rack for towels and clean work clothes in a variety of sizes for the use of those entering the restricted area. A stock of clean shoe covers should also be an hand, and, if possible, a hand-and-foot counter at the exit for making a final check before leaving the facility.

The used clothing will require laundering, preferably through a system separate from that used for laboratory coats and similar items from the nonradioactive areas of the building. The work clothes will frequently receive hard use or may not be worth salvaging if heavily contaminated, so plans must be made for above-average rates of replacement. IAEA No. 22 considers such housekeeping mechanisms in some detail.

Most organizations purchase readily available work clothes of the type used all through industry. Closely woven cotton fabrics are common, probably being replaced in part today by some of the synthetics. Outer disposable garments of paper are obtainable and could be considered for a once-only operation or one that must be carried out very infrequently, following which they would be discarded. The working conditions have to be taken into account and employer-furnished heavy outer clothing for cold weather might be supplied for an environmental survey team, as an example. Many facilities require the wearing of shoe covers ("booties," "shoe-bags") or toe rubbers over regular or safety shoes while the wearer is in a restricted area. The secondary devices are pulled over the shoes when entering, then dropped into a bin or placed on a rack when leaving. The shoe covers may be zippered and reusable types of canvas or heavy cotton, or, probably more frequently, designed for disposal after a single wearing and made of plastic or paper. Toe rubbers (ordinary rainwear variety) may be numbered and assigned to regular employees, or a variety of sizes may be kept in a rack at the entrance of the area for use by all comers. Booties are particularly useful during a tracking incident—contaminated shoes can be covered until there is time to clean them and areas where the presence of floor contamination is uncertain can be entered and monitored without the surveyor himself risking the acquisition of hot shoes.

Gloves are very important items in any radioactivity-handling facility. Light-weight surgeon's latex gloves are used extensively in radiochemistry hood work where possibly contaminated glassware and similar items are handled directly. The same type of glove is also frequently used as a second line of defense when the arms are inserted into the much heavier gloves of a glovebox. The surgeon's gloves can be removed and reworn several times before being discarded if sufficient care is taken. Each glove is turned inside out as it is pulled off the hand and stored on a clean surface until needed again. Some gloveboxes are outfitted with special shelves or with an exterior-opening cubicle into a corner of the box for this purpose. The glove is carefully picked up by the cuff and shaken right side out when it is to be reworn. The practice entails some contamination risk and if any doubt exists as to the cleanliness of the operation the gloves should be immediately discarded after a single wearing.

Glovebox gloves have become a small research area in themselves since they must be impervious to reagents and solvents, difficult to tear and still not so heavy as to cause the wearer to lose all sense of touch. They will not be discussed here, having been ably covered by Barton.[51] Heavy rubber or leather and ordinary work gloves are often used in hot-cell operations for the more strenuous work. Lead-impregnated gloves have been developed for use around x-ray generators. Disposable gloves of thin plastic that are pulled off a roll can be obtained. Gloves may be dispensed with entirely if the hands are covered with an ointment that can be easily washed off at the end of an operation. Such ointments are often used as a secondary protective barrier to be rubbed on the hands before putting on gloves.

The handling of tritium in large quantities poses special problems since the gas diffuses easily through many substances and can be absorbed into the body through the skin. Employees at Mound Laboratory involved in doing maintenance work when tritiated water or oil are present wear five successive layers of gloves.[52]

The gap between the glove cuff and the sleeve of the wearer is a vulnerable spot where contamination can reach the skin. A taping technique can be used if the hazard is likely to be serious. The sleeve is sealed to the wrist with adhesive or masking tape, the glove pulled on and then in turn taped to the sleeve.

IAEA Safety Series No. 22 lists some twenty pre-1967 titles relevant to protective clothing in its bibliography. The protective devices discussed above are primarily for routine everyday use by all workers in a facility. Respirators and their natural extensions, furnished-air masks, and suits may become necessary as additional protection when more hazardous jobs have to be undertaken or when there is an emergency. Respirators are not for routine use, a fact that should be emphasized. Even the lightest-weight device available imposes some physiological stress on the wearer and may seriously impair his vision or movements, thus introducing an additional level of hazard. The basic facility should be modified to improve ventilation or other engineering undertaken in order to eliminate the need for respirators if it is found that these are having to be used almost on an everyday basis.

The U.S. Bureau of Mines started approving self-contained breathing apparatus and gas masks for mine rescue work in 1919 and subsequently added approval schedules for other types of respirators. When the Occupational Safety and Health Administration (OSHA) was formed the agency was given the authority for establishing a generally applicable respirator-approval system for all industry, codified in 1972 in 30CFR11, "Respiratory Protective Devices; Tests for Permissibility; Fees." These regulations require that all respirators be approved by MESA/NIOSH (Mine Enforcement and Safety Administration/ National Institute for Occupational Safety and Health). Certain older Bureau of Mines-approved devices are still acceptable for a few more years, but all newer designs must be certified by MESA/NIOSH before they can be placed upon the market.[53]

Many organizations maintain a group whose responsibility is the maintenance and testing of respirators and similar devices. The lead laboratory in this area for some years has been the Los Alamos Scientific Laboratory which has at various times conducted respirator research, development, and testing for AEC, ERDA, NRC, DOE, NIOSH, and industry. As part of this effort respirator manuals have been prepared for ERDA,[53] NRC[54] and NIOSH.[55] The NRC has also issued a Regulatory Guide[56] and IAEA Safety Series No. 22 contains relevent information and illustrations. The American Standards Institute has prepared a standard.[57]

The basic purpose of such devices is of course to ensure that the wearer will not breathe in hazardous contaminants, radioactive or not, from the air in which he finds himself. These contaminants may be as particulates which must be filtered out unless a separate source of clean air is furnished. Filters are of no use against a hazardous gas, and if such is present the air must be drawn through an absorbent cartridge before it is safe for breathing. A complication here is that different gases will generally require different types of absorbent-bed materials. The filter unit or cartridge may simply be attached to a mouthpiece or bit with subsequent breathing all done through the mouth and with the nostrils closed off with a clip. This type of unit has obvious limitations and is basically for escape under emergency circumstances or for situations that could present possible threat. (The airlines used such devices for some time to solve the problem of getting oxygen to a number of passengers in a hurry if there was a loss of cabin pressure.)

The respirator designs then graduate through the quarter facepiece (nose and mouth covered, with the lower support strap just below the lower lip), half facepiece (nose and mouth covered, support strap under the chin), and full facepiece (eyes also covered by a transparent window—the assault mask familiar to those having had army basic training). Each type may be attached to a filter unit, an absorbent canisher or a combination of both; or air or oxygen may be furnished from a central supply by air hose or from cylinders carried by the wearer.

The next step in the design progression encloses the whole head in a helmet which may have a hood attached that fits down over the body to be tied in at the waist. The final step of course is a helmet attached to a suit that covers the entire body. Air may be supplied by hose as in a diver's suit, or may be carried in cylinders as was done by the astronaut moon walkers.

There is thus a great variety of respiratory protection devices available for different applications. The most common concern in hot-cell work is air-borne particulates, so air-purifying half and full face masks of the filter type are frequently used if the length of time in the hazardous area is to be relatively short in duration and/or the particulate level is not excessive. If a longer stay is necessary or the alpha contamination level higher, recourse to a supplied air suit may be necessary, either of the air-hose type [Figure 5.5(a)] or self-contained. The British[45c] have been particularly active in applying "bubble-suit" or "frog-suit" techniques. The wearer must be carefully observed and emergency

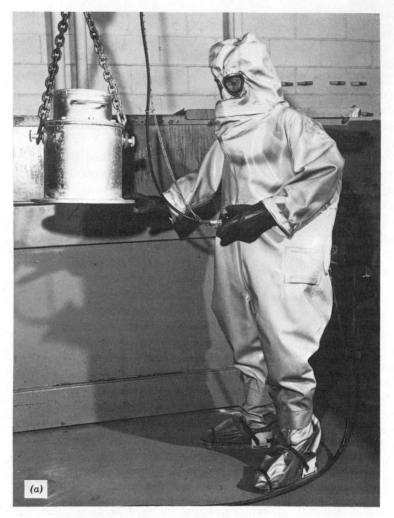

Figure 5.5   (*a*) Frog-suit. (*b*) Newt-suit. (ANL Photos 202-45 and 107-7883.)

rescue plans ready for instant operation if he has difficulties. In the British technique the wearer's suit is throughly scrubbed down by two attendants after his return from the contamination area. The suit is then carefully removed and the operator goes on to take a shower. A German systems cuts down on some of these cleanup activities.[58]

An interesting variation of the furnished air devices is the "newt suit" or "bubble suit" as shown in Figure 5.5(*b*).[59] An alpha-tight barrier is built at the entrance to an area where contamination is expected and a glove-ring large enough for a person to crawl through is installed in this wall. The suit, which is a long tubular sack terminating in a vertical portion in which the operator stands, is fastened to the ring. The bag is pulled inside out into the clean area,

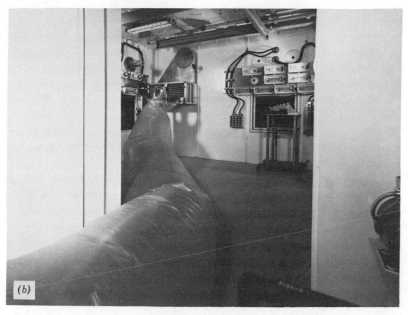

(b)

Figure 5.5　(Continued)

the operator works himself into it and then crawls back through the ring, pulling the bag after him. As with all furnished air suits there are the drawbacks of high noise level from the moving air, some difficulty in breathing normally, possible heat problems and less than ordinary freedom of movement. The devices are not for claustrophobes or those in poor physical condition.

5.4　ORGANIZATION AND ADMINISTRATION

IAEA Safety Series Nos. 1,[1] 38[5] and 39[60] all discuss the matter of organization in facilities handling radioactivity. The responsibilities are defined for the top management, the "Radiological Health and Safety Officer," the individual worker, and the medical services. These matters will not be discussed in great detail here in the same terms, but some comments will be made from a more general point of view.

The size and nature of a particular operation will largely determine the complexity of the administrative organization, but one principle should hold in all cases, that is that the lines of responsibility and authority are very clear to everyone concerned. A divided authority in case of a major emergency could have very serious consequences. A reasonably large organization is assumed here for the purposes of discussion.

IAEA recommends the appointment of one person to have the overall responsibility for radiation safety, the Radiological Health and Safety Officer

(RHSO). Radiation health physics has fortunately become enough of a separate academic area to furnish trained persons up through the Ph.D. level for such positions. Certain responsibilities of the RHSO are obvious from previous sections. The problems are likely to arise in defining other responsibilities in the gray areas between the scientific group leaders, the hot cell supervisors, the chief of the fire department, the head of the medical section and the building maintenance supervisors. Problems do not need to arise if there is good will on all sides, but the potential for trouble is there if the top management is fuzzy in establishing ground rules. The question of authority must also be considered for two different situations, normal operations and emergencies.

Most organizations establish a health physics section (which may go by other names) separate from the groups actually carrying out the programs. The head of this section is equivalent to the IAEA's "radiological health and safety officer." Providing the section with a clearly independent status has the advantage of giving it a quasijudicial position, with the primary responsibility for safety irregardless of program deadlines. The RHSO should have the authority to stop any operation he considers to be unduly hazardous, but with the accompanying responsibility for completely and promptly justifying his action in an immediate hearing before top management. Indiscriminate use of this veto power, in addition to being a constant source of friction, will inevitably destroy any hope of building a good working relationship based on mutual respect between the support groups and those carrying out the programs.

Routine everyday operations are normally not a problem. Appropriate rules are established and the health physicist need only be concerned if it is obvious that the regulations are being ignored. Nonroutine operations, changes brought about by shifts in program and emergencies raise other questions. Burch[45d] argues that the planning, training, emergency drills and other activities associated with nonroutine actions should be the responsibility of the operating group since its members will always have a better detailed understanding of their own area and its current status. Any new plans would be reviewed in advance with the health physicists and other support groups, but implementation of the plans would be by the operating personnel. This is a reasonable point of view, particularly where the support groups have responsibilities in a number of different areas and cannot be expected to know the detailed daily situation in each. Written procedures and drills are essential in the industrial-type operations where large quantities of active material are handled.

Responsibilities within the operating group itself should be equally as clear. Ideally the supervisor or group leader should know exactly where every radioactive source is located and the detailed movements of every person in his area at all times. And, since even supervisors need vacations, go on business trips and become ill, there should be an equally well-informed alternate to take over. This ideal is not easily achieved in a large and busy facility, so some jobs must be delegated. The formal assignments—criticality-control representative, safety-committee chairman, emergency-plan coordinator and special-materials curator (responsible for keeping the books on radioactivity locations and transfers) should all be concentrated near the top, perhaps even in one person, so that the individual having the ultimate responsibility has immediate access to the information he needs.

The rules for records that must be maintained, signs that must be posted, labeling of radioactive samples, training activities, reports that must be made and so on are defined in the various parts of 10CFR discussed in Chapter 2. Many of these activities will be carried out by support groups in a large organization, but some of the chores will inevitably fall on the scientists or operating personnel. One such responsibility which is steadily becoming more visible is that of maintaining better records of the locations of smaller quantities of radioactivity. This is particularly difficult for the research chemist. Any supply of new material usually very rapidly becomes distributed through a number of precipitates, cruds, supernatants and counting plates. The chemist then has the awkward choice between spending all his time in assaying fractions or of getting on with his research.

Another area where the requirements show signs of expansion is that of physical examinations for those working with radioactivity. Most large organizations require such examinations at the times of hiring and termination and schedule others at periodic intervals in between. IAEA Safety Series No. 25[61] deals with the subject of medical supervision of radiation workers. ERDAM Chapter 0528[62] describes the medical programs required of ERDA (now DOE) contractors.

The rules established for everyday operations must be given realistic appraisal in terms of the levels of activity handled and the level of training of the persons affected. Smoking, consumption of food or beverages and putting on cosmetics are obviously risky activities in Zone-3 and -4 areas. Mouth suction on pipettes when handling radioactive solutions is dangerous and supernatants should be drawn off rather than poured off of precipitates. Notebooks, pencils and pens used in the laboratory can be contaminated, so a decision must be made as to whether or not they can be moved to offices if properly surveyed beforehand. Working alone during off-hours is a particularly difficult question in a research laboratory—zeal and enthusiam are very valuable attributes that should not be discouraged, but a hazard is clearly involved. A not entirely satisfactory compromise if the risk is minimal is to require the individual to check at stated intervals with the telephone operator or the health physics representative on night duty. If the risk is at a higher level, a flat rule should be made that no one can work off-hours unless another person is with them or within easy calling distance. All persons entering a building containing radioactivity during off-hours should be required to sign a register giving the times of their arrival and departure.

Every overexposure and contamination incident should be investigated at once and until the occurence is understood. Appropriate disciplinary action should be taken if there has been a flagrant disregard of the rules. A formal committee of inquiry should be established if the incident was nontrivial, or if a number of events occur within a short period of time. These investigative committees consume an inordinate amount of valuable time, but have a power-ful educational value, particularly for the more chronic offenders.

A large group involved in handling radioactivity should have its own safety committee, given enough status to be certain that its recommendations will be carefully considered.

A representative from smaller groups should always be on the safety com-mittee of a larger organization if such groups are the only ones working with active material. Handbooks spelling out the rules for a controlled area and issued to each employee in that area are excellent if there are mechanisms for making certain that the documents are read and understood. The rules of the game can also be posted in the working areas, but this must be done judi-ciously. If entire walls are plastered with paper there is a definite chance that no one will bother to read any new posting out of sheer boredom.

REFERENCES

1 Chapter 2, Ref. 39.

2 Anon., "A Bubble Chamber in Aspic." Science News, Feb. 10, 1979, p. 89.

3 Gerhart Friedlander, J. W. Kennedy and J. M. Miller, *Nuclear and Radiochemistry*, Second Edition. Wiley, New York (1955).

4 ICRU (Chap. 2, Ref. 24), "Radiation Protection Instrumentation and Its Application." ICRU Report 20 (1971).

5 Chapter 2, Ref. 20.

6 F. H. Attix and W. C. Roesch, Eds., *Radiation Dosimetry, Instrumentation* Volume II, Second Edition. Academic Press, New York (1966).

7 K. Z. Morgan, "Techniques for Personnel Monitoring and Surveying." In A. H. Snell, Ed., *Nuclear Instruments and Their Uses*, Volume I. Wiley, New York (1962), pp. 391–469.

8 Jacob Shapiro, *Radiation Protection. A Guide for Scientists and Physicians*. Harvard University Press, Cambridge, Mass. (1972)

9 Chapter 1, Ref. 12.

10 U.S. Environmental Protection Agency (EPA), "Environmental Radioactivity Surveillance Guide." USEPA Report ORP/SID 72-2 (1972).

11 Anon., "Instrumentation for Environmental Monitoring, Volume 3, Radiation." USAEC Re-port LBL-1, Vol. 3 (1972); Addition No. 1 to Vol. 3 (1973).

12 ANS (Chap. 1, Ref. 12), "American National Standard Criticality Accident Alarm System." ANS Report ANSI N.16.2-1969 (1969).

13 R. D. Friesen, "Fast Neutron Detector for Use as a Criticality Monitor." *IEEE Trans. Nucl. Sci.*, **NS-24**(1), 657–658 (1976).

14 Chapter 3, Ref. 42.

15 Eberline Instrument Corp., P. O. Box 2108, Airport Road, Santa Fe, N.M. 87501.

16 J. G. Ello, "Early Detection of Contamination by the Use of Gas Proportional Monitors." *Health Phys.*, **9**, 653–663 (1963).

17 R. W. Fergus, "A Practical Hand-Shoe Monitor." *ibid.*, **19**, 307–310 (1970).

18 R. J. Walker and W. J. Roach, "An Air Proportional Alpha Hand and Shoe Counter." USAEC Report UCRL 10373 (1962).

19 J. G. Ello, W. H. Smith and J. H. Pingel, "Gas Proportional Alpha, Beta-Gamma Continuous Air Monitor." *Health Phys.*, **11**, 773-778 (1965).

20 Victoreen, Inc., 10101 Woodland Ave, Cleveland, Ohio 44104.

21 NCRP (Chap. 2, Ref. 7), "Radiological Monitoring Methods and Instruments." NCRP Report No. 10 (1952).

22 D. J. Rees, *Health Physics. Principles of Radiation Protection.* M.I.T. Press, Cambridge, Mass. (1967).

23 H. J. Moe, S. R. Lasuk and M. C. Schumacher, "Radiation Safety Technician's Training Course." USAEC Report ANL 7291 (1966).

24 Chapter 2, Ref. 40.

25 Chapter 3, Refs. 54-57.

26 R. Brown, H. E. Meyer, B. Robinson, and W. E. Sheehan, "Ruggedized Ultrasensitive Field Air Sampler for Differentially Determining Tritium Oxide and Gas in Ambient Air Samples." Proceedings of the Third International Congress of the International Radiation Protection Association, Washington, D. C., Feb., 1974. pp. 1434-1339.

27 M. N. Dixon, C. F. Holoway, B. L. Houser and D. G. Jacobs, "Indexed Bibliography on Tritium: Its Sources and Projections, Measurement and Monitoring Techniques, Health Physics Aspects and Waste Management." USERDA Report ORNL 5057 (1975).

28 J. G. Ello, "Modification of the Eberline PAC-3G and FM 3G Survey Meters to Detect Low-Energy Beta Activity." *Health Phys.*, **12**, 1139-1141 (1966).

29 USERDA, "Proceedings of the Conference on Decontamination and Decommissioning (D & D) of ERDA Facilities. CONF 750827, Idaho Falls, Idaho, Aug. 19-21, 1975.

30 H. J. Blythe, A. Catherall, A. Cook and H. Wells, Eds., *Proceedings of the First International Symposium on the Decontamination of Nuclear Installations, Harwell, England, May 4-6, 1966.* Cambridge University Press, New York (1967).

31 G. R. Bainbridge, P. A. Bonhote, G. H. Daly, E. Detilleux and H. Krause, "Decommissioning of Nuclear Facilities, A Review of Status." *At. Energy Rev.*, **12** (1), 146-160 (1974).

32 J. A. Ayres, Ed., *Decontamination of Nuclear Reactors and Equipment.* Ronald Press, New York (1970; USAEC Report TID 25450 (1970).

33 G. D. Schmidt, "Limits for Radioactive Surface Contamination." In N. V. Steere, Ed., *Handbook of Laboratory Safety*, Second Edition. Chemical Rubber Co., Cleveland, Ohio (1971), pp. 477-481.

34 H. J. Dunster, "Surface Contamination Measurements as an Index of Control of Radioactive Materials." *Health Phys.*, **8**, 353-356 (1962).

35 NBS (Chap. 2, Ref. 18), "Safe Handling of Radioactive Materials." NBS Handbook 92 (1964).

36 NRC (Chap. 2, Ref. 14), "Termination of Operating Licenses for Nuclear Reactors." NRC Reg. Guide 1.86 (6-74) (1974).

37 NRC, "Guidelines for Decontamination of Facilities and Equipment Prior to Release for Unrestricted Use or Termination of Licenses for Byproduct, Source or Special Nuclear Material." Draft document received June, 1979.

38 W. H. Schulze, USNRC. Personal Communication.

39 D. G. Stevenson, "Radiological Decontamination. Theoretical and Practical Aspects." *Research (Lond.)*, **13**, 383-389 (1960).

40 W. E. Sande, H. D. Freeman, M. S. Hanson and R. I. McKeever, "Decontamination and Decommissioning of Nuclear Facilities, A Literature Search." USERDA Report BNWL 1917 (1975).

41 H. Phillip and E. Wichmann, Comps., "Bibliographies in Nuclear Science and Technology, Section 13, Decontamination, Number 10." Report AED-C-13-10 (1974).

42 H. D. Raleigh, "Radioactive Decontamination, A Literature Search." USAEC Report TID 3535 (Suppl. 1) (1965).

43 G. A. West and C. D. Watson, "Decontamination Testing of Highly Contaminated Coatings and Gaskets." USAEC Report ORNL 2811 (1960).

44 Decontamination Section, Kernforschungzentrum, Karlsruhe, West Germany, Abstracts in INIS Atomindex, Nos. 222296, 237167, 242422, 299238, 303608, 306590, 306591, 324903, 347255 (1976-1977).

45 (Grenoble Hot Lab Meeting, Chap. 4, Ref. 6): (a) C. D. Watson, E. A. Berreth, J. E. Dummer, R. K. Fuller and R. F. Stearns, "Policy and Procedures Used in the United States for Decontaminating Hot Cells," pp. 677-698. [Also published as USAEC Report ORNL-P-1154 (1965)]; (b) K. R. Ferguson, "United States Practice in Control of Contamination," pp. 889-917; (c) H. Wells, "Decontamination Procedures for High Activity Cells at A. E. R. E. Harwell," pp. 719-732; (d) W. D. Burch, "Organization of the Intervention on Hot Facilities Mishaps, Maintenance, Dismantling," pp. 559-566.

46 K. J. Schneider and C. E. Jenkins, Coords., "Technology, Safety and Costs of Decommissioning a Reference Nuclear Fuel Reprocessing Plant." USNRC Report NUREG 0278 (1977. In two volumes.

47 Turco Products, 24600 Main St., P. O. Box 6200, Carson, Calif. 90749.

48 J. W. Schulte, F. J. Fitzgibbon and D. S. Shaffer, "The Use of Solvent-Soluble Films in Decontamination." (Chap. 4, Ref 2h), pp. 332-336.

49 G. A. West and C. D. Watson, "Radiation Damage and Decontamination Evaluation of Protective Coatings and Other Materials for Hot-Laboratory and Fuel-Processing Facilities." USAEC Report ORNL 3589 (1965).

50 IAEA (Chap. 1, Ref. 2), "Respirators and Protective Clothing." IAEA Safety Series No. 22 (1967).

51 Chapter 3, Ref. 59.

52 T. B. Rhinehammer and P. H. Lamberger, "Selected Techniques for the Control and Handling of Tritium." (Chap. 4, Ref. 2w) pp. 94-99.

53 D. D. Douglas, A. L. Hack, B. J. Held and W. H. Revoir, "Energy Research and Development Administration Division of Safety, Standards and Compliance Respirator Manual." US-ERDA Report LA-6370M (1976).

54 J. L. Caplin, B. J. Held and R. J. Caplin, "Manual of Respiratory Protection against Airborne Radioactive Materials." USNRC Report NUREG 0041 (Final) (1976).

55 J. S. Pritchard, "Guide to Industrial Respiratory Protection." USERDA Report LA 6671-M (1977).

56 NRC (Chap. 2, Ref. 14), Acceptable Programs for Respiratory Protection." NRC Reg. Guide 8.15 (10-76) (1976).

57 ANSI (Chap. 3, Ref. 13), "American National Standard Practices for Respiratory Protection." ANSI Standard Z 88.2 (1969).

58 I. Csiba, "New Frogman Technique." (Chap. 4, Ref. 2i), pp. 331-338.

59 R. C. Goertz, K. R. Ferguson, J. F. Lindberg, D. P. Mingesz, R. A. Blesch, C. W. Potts, J. H. Grimson, G. A. Forster, F. L. Brown and J. L. Armstrong, "The ANL Alpha-Gamma Hot Cell—Its Design Philosophy and Components." (Chap. 4, Ref. 21), pp. 285-306.

60 IAEA (Chap. 1, Ref. 2), "Safe Handling of Plutonium." IAEA Safety Series No. 39 (1974).

61 IAEA, "Medical Supervision of Radiation Workers." IAEA Safety Series No. 25 (1968).

62 ERDAM (Chap. 2, Ref. 44), "ERDA Contractor Occupational Medical Program." ERDAM Chapter 0538 (8/75); ERDAM Appendix 0528A (Handbook) (8/75).

chapter 6

Special Equipment

"Special equipment" in this case refers to items that have been developed or modified specifically for the handling of or undertaking of various measurements on highly radioactive materials. The emphasis will be primarily on bench-scale operations and with a strong bias towards those involving chemistry.

6.1 GENERAL PRINCIPLES

The importance of constructing or modifying equipment so that as much as possible of the device is outside a hood, glovebox or cell has already been emphasized. The parts inside the enclosure are not only subject to contamination and radiation damage but also in many cases to corrosive fumes. Maintenance is also considerably simplified if the more critical working parts do not have to be handled through heavy gloves or by manipulator. Everything of course cannot be left outside, but those elements of the device that must be in the interior should be kept to a minimum and given strippable coatings, local shielding and as much other protection as can be devised without interfering with operations or the equipment's function. The same principle can often be applied to cold reagents needed in chemical work, with stock solutions on the outside being introduced as needed through tubing in wall penetrations to the working area. Traps or check valves may be necessary if there is any chance of back-up.

The equipment for hot cells and gloveboxes in particular should be rugged and simple in operation and extremely reliable in use, although its useful life may be deliberately made short if it is an expendable item. Since it is often not economically worthwhile to decontaminate and salvage some items (support racks, simple glassware, tubing, etc.), efforts should be made to use inexpensive materials when feasible so that they can be discarded after use. It is generally less costly to use commercially available equipment, although such items may have to be modified to a greater or lesser degree for remote operation. Equipment must be chosen that can be remotely maintained, or, if necessary, re-motely replaced with the available manipulation. If the item is such that remote replacement is impossible, an alternate method of carrying out its function should be available in case of failure. Corrosion- and radiation-resistant materials of construction for the exposed parts of the equipment should be chosen to the maximum degree possible. Devices or instruments with versatility allowing performance of more than one function or type of analysis have obvious advantages.

Most of the equipment used and handled in a radioactive hood need be little different from that used on an open bench (if the contamination problem is ignored), although the available space may be more cramped and awkward to use with tall items such as long ion-exchange columns. The space available in a glovebox is usually even more restricted, but here the chief problems are the loss of sensitivity of touch and movement through the heavy gloves and the fact that the operator's arms are fixed in position by the glove ports. The use of master-slave manipulators in a hot cell generally improves the space-limitation problem, but offsetting this improvement are the loss of ability to distinguish directly fine detail through the thick viewing windows, the fact the manipulator hand is limited to two fingers that operate only to open and close against each other, and the loss in the ability to rapidly change direction of movement in the same manner characteristic of the human arm and hand. Manipulator turning of a crank is an awkward process.

Extensive "dry runs" on cold materials should be made in advance to test equipment behavior under conditions simulating those of the proposed operation as much as possible. Much preplanning is also indicated to be certain that the effects of any failure can somehow be nullified by remote means if it should become necessary. If equipment is to be salvaged at the end of the experiment, methods for accomplishing this should be worked out in advance.

Hot-cell operations are inevitably slower and often more tedious to carry out than are the same procedures used under normal conditions. It is sometimes possible to reduce the radiation level to the point where cell operations are not necessary for the complete procedure, a technique used in several different ways in the analytical laboratories at Karlsruhe.[1a] There are many other arguments that can be advanced for the use of micro- or semimicromethods in radiochemistry (increased speed, fewer radiation effects, conservation of materials) and in the early days of the Manhattan Project nonavailability of materials in quantity made the use of ultramicromethods a necessity, a situation still true for research on the very heavy elements. Monk and Herrington[2] point out that many radiochemical procedures require the use of an inactive carrier, generally of the order of 10 to 50 mg. This much material sometimes leads to counting uncertainities due to self-absorption in the sample. Monk and Herrington attempt to reduce the amount of required carrier to the 0.5–1.0-mg range by going to micromethods in determining the final chemical yield. They describe some of these methods and some of the special equipment they have designed for applying them. The text by Kirk[3] is an excellent source of ultra-microtechniques and associated equipment.

6.2 CHEMICAL OPERATIONS

Laboratory-scale operations in cells and gloveboxes generally fall into one of two categories—repetitive and nonrepetitive. The first of these is typified by the support laboratory for a process plant or similar facility where large numbers of samples are produced, generally rather uniform in nature and requiring analysis for a well-defined and relatively unchanging group of elements and radioactivities. Chemistry in gloveboxes or cells prior to fabrication of reactor-fuel elements is somewhat similar since, while very large quantities of material are processed, the chemistry used follows a set of standard procedures on a routine basis. The nonrepetitive category encompasses research and development activities. A variety of different techniques may be needed, sometimes only used once. A much smaller number of samples require analysis, but some of these analyses may in themselves require their own research and development program for solution. The repetitive, continuing type of activity will require equipment of high reliability and comparatively long working life, and the development of such items can justify considerable expense and time. The nonrepetitive effort can afford more improvisation (outside of equipment peculiar to the purpose of the research) since many items used will be discarded at the end of the experiment. Literature references to equipment for the more production-oriented remote chemistry facilities are accordingly more abundant.

The predecessor chapter[4] to this volume lists a large number of references to earlier sources of information on chemical equipment developed for work with radioactive materials. Many of these reports or papers are now difficult to access and will not be recited. More accessible older sources containing much still useful material are Garden and Nielsen,[5] Metz and Waterbury,[6] Higgins and Crane,[7] Stang,[8] Dykes et al.,[9] Shank et al.,[10] Lamb,[11] Hawkins,[12] Ruehle[13] and Miller and Kinderman.[14]

A complete 1970 volume[15] was devoted entirely to remote analytical chemistry and separate chapters prepared by a number of specialists. Barton's recent chapter[16] on glovebox techniques contains sections on analytical work and on physical, chemical and metallurgical operations. Well over 1000 citations to literature sources are given between these two surveys. No other comprehensive reviews appear to have been prepared in recent years and individual papers limited entirely to equipment descriptions are few over the same period. However, as would be expected, the information that is available[1a,1b] would indicate an increasing use of computers in the high-sample-volume facilities for both data accumulation and evaluation and to a lesser extent for direct operation of analytical procedures. The latter area will obviously grow in importance for hot-cell work.

Only a few of the more common chemical techniques will be touched upon here; primarily in terms of their use in hot cells since glovebox and hood applications are generally more straightforward.

Remote weighing of active material has been somewhat simplified in recent years by the advent of the single-pan balance, built-in weights, automatic taring, digital readouts and now printout of data. Weighings comparable in accuracy to those made in an ordinary laboratory are however still difficult to make in a glovebox or cell. Air currents due to the forced ventilation (and in some cases, convection currents generated by the heat of the sample itself), vibration, corrosive fumes, radiation damage to optical and electronic components, and static buildup in dry atmospheres can all add complication to the already awkward problem of handling a very sensitive instrument under difficult conditions.

A commercial balance can be used directly in a glovebox without modification, although routine maintenance is thereby essentially abandoned and the working lifetime of the instrument will be probably short. An ordinary balance in a hot cell will almost certainly require some modification, primarily in replacement of the operating knobs with types suitable for manipulator handling. The door to the cabinet containing the weighing pan may also have to be modified for the same reason, and provision made to be certain that any readout devices can be clearly seen through the viewing window. The difficulty of handling small weights by manipulator makes the single-pan balance the obvious choice. The problems of maintenance and calibration are even more severe than in the glovebox case.

If there is sufficient space at the side or on top of the cell or box, a more satisfactory arrangement is to leave only the weighing pan in the contaminated area as shown in Figures 6.1(*a*) and 6.1(*b*) which picture installations at the Idaho Chemical Processing Plant.[15] In Figure 6.1(*a*) a shielded cubicle has been

Figure 6.1 (*a*) Analytical balance at end of line of cells. (From Ref. 15a.) (*b*) Glovebox analytical balance. (From Ref. 15a.) (*c*) Commercially available remotely controlled balance (Courtesy Torsion Balance Co., Ref. 17.)

built alongside a row of cells and the balance placed on top. The pan is suspended on a long wire and is within the shield. Some method of course must be provided for moving samples into the cubicle. In the case shown, the pan is unhooked, moved by a small trolley into the cell complex, loaded, and then moved back into the cubicle for reattachment to the balance. In the arrangement shown in Figure 6.1(*b*) the balance is on top rather than at the side of the enclosure. Sample handling is simplified, although the raised position makes the balance somewhat awkward to operate. A small open penetration between the hot and cold areas is of course required in both arrangements for passage of the suspension wire to the pan. The negative air pressure inside the enclosure aids in restricting contamination spread, but it may become necessary to limit this even more by placing the balance cabinet within its own separate enclosure. The pan in the interior may also have to be enclosed, as well as the

Figure 6.1 (Continued)

suspension wire [as shown in Figure 6.1(*b*)] to eliminate vibration problems due to air currents.

Various remotely operated balances have been produced commercially, one of which is shown in Figure 6.1(*c*). Operating knobs and door openers have been designed for manipulator handling and fabrication materials were chosen for corrosion and radiation resistance. Weights down to 0.1 g are adjusted with the manipulators. Weights adjustments and readout below 0.1 g are made with a potentiometer in the electronics cabinet located outside the cell.

The weighing installation at HFEF-N[18] has a capacity of 4000 g and a sensitivity of 0.02 g and delivers a printed readout of the weight to the operator. This device, for weighing fuel elements, is of course approaching the industrial scale of operations.

Pipetting of accurately known volumes of solutions, like weighing, is one of the fundamental chemical operations. Pipetting is unusually important in radiochemical work where many analyses are made by radiometric techniques which depend on taking aliquots small enough to allow them to be removed from a contaminated area to an external counter. Micropipets are accordingly used extensively. These are a commonplace today, primarily due to such nuclear applications.

Handling pipets in the 1 μl–5 ml range in a radiochemistry hood or glovebox is reasonably straightforward. One end of a one-hole rubber stopper is fitted over the glass tip of a syringe of appropriate size, and the pipette inserted

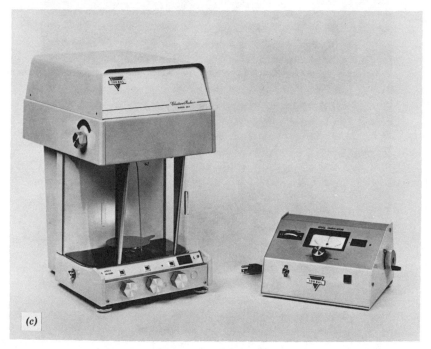

(c)

Figure 6.1 (Continued)

in the other. (The stems of ordinary pipettes in the 1–5-ml range can be pulled down in a flame to a diameter suitable for insertion into the stopper.) Drawing on the syringe fills the pipet, the meniscus level is adjusted to the line, the external pipet tip wiped off, and the solution expelled when desired by means of the syringe plunger. Small (1–10-μl) pipets that fill themselves by capillary action are available, and eliminate the problem of adjusting the meniscus. They however have to be emptied by syringe action. Pipets much larger than 5 ml are awkward to handle with a syringe, but can be filled and emptied by means of a rubber squeeze bulb. Pipet-filling bulbs are available commercially having a side tube fitted into the stem of the bulb. This tube is covered by a finger in use and allows manipulation of the bulb pressure to assist in adjusting the meniscus level.

The syringe technique has been adapted for remote pipetting in a hot cell. In the simplest form a small-diameter metal or plastic tube through the shielding wall connects a syringe with a stopper or other device into which a pipet can be inserted. Holes may be drilled into both ends of the syringe plunger, the one on top being covered by the operator's finger. This again allows somewhat more sensitive control of the pressure in the connecting line. Adjusting the solution meniscus can be a frustrating experience with such simple air-buffer designs if the tube through the wall is long. Much more satisfactory are devices such as the Berkeley hydraulic-diaphragm micropipet[19] diagramed in Figure 6.2. Turning the screw in the syringe control on the left moves a piston against or away from an ethylene glycol reservoir that fills the system over to a vinyl

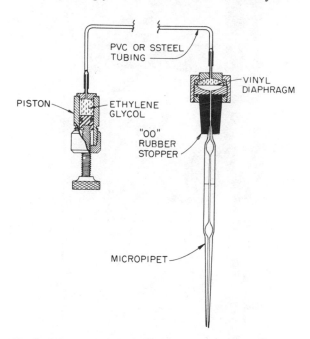

Figure 6.2 Berkeley remotely controlled micropipet. (From Refs. 19 and 15b.)

diaghragm in the filling unit in the cell. Pressure changes on this diaphragm cause the solution to rise or fall in the pipet. This system allows good control of the meniscus level. Units based on the same principle have also been designed for macropipetting. In this case the pipets are custom-made and connected to the pressure-control system by semi-ball-joints as shown in Figure 6.3. (The high loop in in the capillary draw tube keeps solution from draining back from the pipet bulb when contact is broken with the source liquid at the pipet tip.) The same design is used for devices used in transferring liquids from one container to another. In this latter case the tube carrying the male semi-ball-joint and connecting to a pressure-control device outside the cell is fitted with a

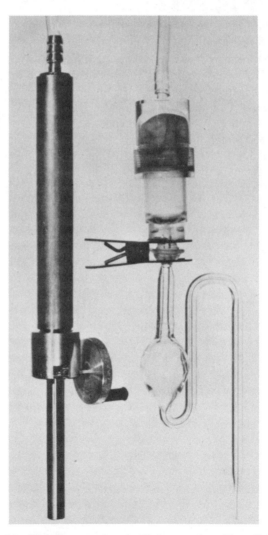

Figure 6.3   Berkeley remotely controlled macropipet. (From Ref. 15b.)

plastic collar of a design that allows the entire unit to be moved from place to place within the cell by the manipulators.

Ertel describes[1a] a unit for 50 μl-1 ml pipetting used at Karlsruhe. A piston microburet is used rather than a syringe to change pressure, and an expansion chamber in the cell replaces the diaphragm. The buret, through-the-wall tubing, and expansion chamber (which is connected to the pipet insertion device by stainless-steel microtubing) are filled with mercury. Disposable plastic pipets are used. The pipetting position remains fixed in location and solutions for aliquoting are brought to it by the manipulators.

An obvious problem in "manual" remote pipetting of the type described is that of seeing the meniscus through the shielding window, particularly when using micropipets. Direct viewing is often possible if the lighting is good, the window clear and the shielding wall not too thick. Binoculars are sometimes a help. Recourse to a periscope installation may be necessary if the distance between the operator and the working area is too great. The self-filling capillary pipets may also be of benefit in such a situation.

Relatively simple pipetting devices such as described are effective tools in research where the aliquoting demands are highly variable, but limited in number. The technique is however slow and thus not very satisfactory when large volumes of samples have to be processed. A number of more elaborate devices have been developed to solve this problem and are individually reviewed by Maddox.[15b] These automatic pipetters are all motor driven, often controlled by electronic servomechanisms, and can become quite complicated in design. They all have the advantage that the tubing penetrations through the wall required by the simpler devices are replaced by electrical connections.

Remote titrimetry is routinely carried out in the hot cells of analytical control groups such as those at Oak Ridge, the Idaho Chemical Processing Plant, and similar organizations in this country and abroad, but is very rarely necessary in most other radiochemical work. This is fortunate since the devices required are of necessity complex in design and usually expensive to construct. The subject of remote titrimetry has been thoroughly reviewed by Thomason[15c] and by Shults.[15d] Commercially available automatic titrators have been successfully modified for remote operation in some cases. The references cited describe potentiometric, colorimetric, amperometric, conductometric, photometric, radiometric, and thermometric titrations. The analyses for uranium and plutonium by titrimetric methods have received extensive attention.

Sampling of liquid solutions was discussed to a limited extent under pipetting. Other approaches are possible, particularly in process work when it is desired to transfer enough liquid from the cell back to the laboratory for several different analyses. One method is to place two hypodermic needles in a fixed holder as shown in Figure 6.4. The lower needle is connected by tubing to the solution to be sampled, the other to a vacuum through an intervening waste trap or back to the vessel being sampled. The sample bottle, closed by a thin rubber diaphragm, is raised against the needles until they penetrate through the cap and into the bottle. Application of a mild vacuum fills the container. The liquid flow is then valved off and excess liquid above the tip of the higher needle is expelled, leaving a sample of standardized volume. A variation[20] uses only one double-pointed needle, one end being introduced into the liquid to be sampled. The other end is then raised to penetrate the diaphragm cap of an inverted sample bottle. The previously evacuated container then fills itself by suction action.

Sampling of solids under any conditions always involves the question of obtaining a representative sample, an uncertainity whose resolution is not helped in remote operations. A number of small samples taken at various points on the solid may be the only way of gaining reasonable assurance that the final analyses represent the true composition of the material.

Many different techniques have been used in hot-cell work for solid sampling, the specific method determined by the physical properties and the shape and form of the material. Jones[21] (as quoted by Lamb[15e],) says, "Samples of material too thin to be machined . . . are taken by clipping, shearing or punching . . .. Rods, bars, plates, shapes, tubes, etc. . . . are sampled by milling the entire cross section, by drilling through the material at several points, or by sawing out pieces." Electrolytic dissolution can be used in some cases. An

Figure 6.4   Device for sampling multiple sets of process solutions. (From Ref. 15e.)

obvious problem in using machining techniques is that of avoiding contamination of the sample with metal from the machine itself.

Blending of crystalline, powdered or granular material in order to obtain a representative sample is also a problem. Manipulator accomplishment of the chemist's traditional method of coning and quartering is simply not practicable. Mixing mills, Jones' riffles and blenders of various types have been used, but all present the same problem of adequate cleaning between applications in order to avoid cross contamination of samples.[15e]

The separation and purification of inorganic ions by solvent extraction or ion exchange have been widely used techniques in nuclear work since the early days of the Mahattan Project. The scale of application has varied from handling an almost countable number of atoms in ultraheavy-element research to full industrial production of tons of material in reactor-fuel reprocessing. Morrow[15f] has reviewed laboratory-scale applications.

The ideal solvent-extraction system would be one in which the distribution ratios of the species of interest and its associated impurities were such as to produce a 100% extraction of the former along with complete decontamination from the latter in one contact between the aqueous and the organic phases. Such an ideal does not exist, of course, but there are some systems giving a reasonable approximation. The equipment needed in these few cases can be very simple—the two phases are mixed in a container, allowed to separate, and the solvent fraction containing the product of interest drawn off by transfer pipet, or even an eye dropper if the scale is appropriate. Such single-stage situations are not however the norm. Additional extractions may be required to improve the yield, and the organic phase may need scrubbing with a suitable aqueous solution in order to obtain better decontamination. A sequence of single operations of this type becomes difficult to accomplish by remote means. Recourse is generally made to miniature mixer-settlers, pulse extraction columns, or extraction chromatography for such multistage extractions.

In the mixer-settler the organic phase is transferred mechanically in one direction through a series of compartments, while the aqueous phase similarly moves through the same compartments in a countercurrent direction. The pulsed extraction column operates in a somewhat analogous manner with the lighter organic phase moving upward in a column countercurrent to the heavier aqueous phase moving downward. These two techniques are usually applied when somewhat larger quantities of material are involved and when there is a relatively substantial difference in the distribution ratios (K_D's) of the product and the impurities between the two phases. Such K_D ratios are not always that favorable, particularly for separation of chemically very similar elements such as the rare earths and the + 3 actinides. Ion exchange or extraction chromatography must be used for such difficult situations if clean separations are a necessity.

In extraction (reverse-phase) chromatography the aqueous feed is loaded onto the top of a column filled with a hydrophilic inorganic substrate. The solvent is then fed into the column to elute the products downward, those with the higher K_D's moving more rapidly in a manner analogous to ion exchange. With care, very short columns can be prepared to provide the equivalent of hundreds or even thousands of single-stage extractions. Horwitz and his collaborators[22,23] have developed elegant extraction chromatographic techniques permitting essentially complete separations within minutes of element pairs extremely difficult to separate from each other by other methods. The chief drawbacks of the reverse-phase method are the extreme care needed to prepare a good column bed and the somewhat limited amount of material that can be loaded onto the substrate. This last objection is being eased by the development of new bed materials.

Ion-exchange techniques have been used in several ways in nuclear work. In the first, the feed solution is adjusted in composition so that the product of interest is preferentially strongly absorbed by the resin, whereas the impurities are not. The absorbed product at the top of the column is then washed with impurity-removing solutions, then finally stripped off for recovery by changing the elutriant to one in which the product affinity for the resin in much smaller. This is a purification procedure. The second ion-exchange approach is for element separations. The product mixture is loaded onto the top of the resin column and selectively eluted off, usually with a solution whose anions complex the elements in the feed to a varying degree. The elutriant from the column is collected in a series of containers to obtain fractions containing individual elements from the original mixture. This chromatographic technique is difficult to apply remotely on a large scale, so is primarily a research tool. (Large-scale industrial applications have been made in separating nonactive rare earths.)

A wide variety of equipment has been developed for both remote ion exchange and extraction chromatographic applications. Morrow describes some of these items as well as instrumentation for automatic collection of elutriant fractions and determination of their individual volumes and radioactivity levels.

A high-sensitivity spectrograph is a large instrument whose operation would be difficult by remote means. The sample sizes it requires for analysis are however fortunately very small. It is thus frequently possible to bring small aliquots out of a cell to an externally located spectrograph having only the electrode compartment enclosed and separately ventilated; and perhaps with some local shielding. The spectrograph is usually used as an analytical instrument to determine impurity levels in a product of interest. When the sample is so active as to make removal of even a small fraction difficult, ion-exchange and solvent extraction techniques have been developed for concentrating the impurities, leaving the bulk of the activity behind. The impurities fraction is then analyzed. Spectrographic analysis of radioactive materials has been reviewed by Gaddy.[15g]

Spectrophotometers and colorimeters of different kinds find considerable use in nuclear facilities, again usually for analytical applications. Dykes and his associates[15h] describe various arrangements for adapting such instruments for

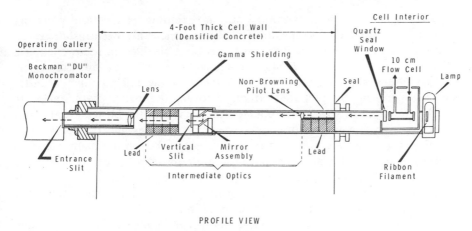

PROFILE VIEW

Figure 6.5   A remotely controlled shielded spectrophotometer. (From Refs. 24 and 15h.)

remote work. As discussed in Section 6.1, the primary aim is invariably to keep as much of the instrument as possible outside of the hot area. Figure 6.5 shows an application of this principle described by Upson and Wheelwright[24] for a Beckman DU spectrophotometer. The light beam passing through the shielding wall is offset by mirrors to allow placement of lead shielding in the through tube.

Dykes and his associates also discuss adaptation of modified commercial atomic absorption and flame emission spectrometers for remote work.

6.3   NONCHEMICAL OPERATIONS

Fundamental metallurgical studies and empirical postirradiation examinations of materials exposed to radiation are vital to the development of nuclear power and to the construction of large research accelerators. There however does not appear to have been any summary review made specifically for surveying the equipment needed for these activities. Numerous pertinent individual papers are scattered through the Hot Lab Group Proceedings,[25] undoubtedly the best source for detail. Some information can also be found in the *Reactor*[26] and *Plutonium*[27] *Handbooks* and in Barton's review, previously cited. The proceedings of a series of international meetings on the subject of plutonium (now broadened to include the other actinides) contain some equipment descriptions, but these are usually incidental to the reporting of research results. Meetings were held in 1955,[28] 1960,[29] 1965,[30] 1970[31] and 1975[32] The emphasis at the first of these conferences was more strongly on facilities and equipment than has been the case in the later symposia.

A modern metallograph is an expensive and complex instrument requiring the maximum amount of protection in its installation. This problem has been met in remote facilities in various ways, all of which are designed to keep the preparatory activities such as sample cutting, grinding, etching and polishing separate from the metallograph itself. The instrument has been installed in a separate cell of its own in a few cases, has been shielded in its own cubicle within a cell or housed in a "blister," a shielded cubicle built against the external side of the shielding wall, that is, outside of the cell proper.[33] Transfer devices for moving the prepared samples between the metallograph enclosure and the preparation area are of course necessary, and the blister installations are accordingly provided with suitable penetration holes into the cell proper. HFEF-N specimens are prepared in a containment box in the main cell, then transferred to a smaller shielded cell in a separate room by means of a pneumatic tube.[34] The facility at General Atomics in San Diego slides the stage portion of the metallograph into the working cell as needed. When not in use the stage is retracted into a shielded opening in the wall.[33]

Modern metallurgy employs a large number of highly sophisticated examination instruments in addition to the metallograph. At Oak Ridge two metallographs and a microhardness device are housed in separate blisters at the side of the metallography cell. In addition, a sample transfer device can move specimens to a shielded, manipulator-equipped cubicle on the floor above. The sample can then be routed from this cubicle to a shielded x-ray diffractometer or to an electron microprobe. The cubicle can also be used for preparing plastic replicas for electron microscope examination, for the making of alpha and beta-gamma autoradiographs, for vacuum deposition of metallic or carbon coatings and for cutting of ultrathin sections of radioactive materials for scanning electron microscopy.[35] Most of the major nuclear-fuel examination and metallurgy facilities in this country and abroad have a similar broad range of capabilities.

The larger-scale examination of irradiated reactor fuels and construction materials is beyond the scope of this volume. The Hot Lab Group Proceedings describe much relevant remote equipment in detail—tensilometers, nondestructive testing equipment such as eddy-current devices, gamma-scanners, laser profilometers, neutron radiography installations, and so on.

REFERENCES

1 (Finland Hot Lab Meeting, Chap. 4, Ref. 8): (a) D. Ertel, "Equipment and Instrumentation of a Laboratory for Purex Process Analytical Chemistry," pp. 159-164; (b) P. Groll, "Automation of Analytical Processes," pp. 81-86.

2 R. G. Monk and J. Herrington, "The Use of Microchemical Methods in Radiochemical Analysis." *Anal. Chim.* Acta, **24**, 481-492 (1961).

3 P. L. Kirk, *Quantitative Ultramicroanalysis.* Wiley, New York (1950).

4 Chapter 3, Ref. 2.

5 N. B. Garden and E. Nielsen, "Equipment for High-Level Radiochemical Processes." *Ann. Rev. Nucl. Sci.*, **7**, 47-62 (1957).

6 C. F. Metz and G. R. Waterbury, "The Transuranium Actinide Elements." In I. M. Kolthoff and P. J. Elving, Eds., *Treatise on Analytical Chemistry,* Part II, Vol. 9. Interscience, New York (1962), pp. 189-440.

7 G. H. Higgins and W. W. T. Crane, "The Production and Chemical Isolation of Curium-242 in Thousand Curie Quatities." Proceedings of the Second International Conference *on Peaceful Uses of Atomic Energy, Geneva, 1958,* **17,** 245-251 (1958). A/CONF 15/P1883, United Nations, New York.

8 Chapter 3, Ref. 23.

9 F. W. Dykes, R. D. Fletcher, E. H. Turk, J. E. Rein and R. C. Shank, "Laboratory for Remote Analysis of Highly Radioactive Samples." *Anal. Chem.,* **28,** 1084-1091 (1956).

10 R. C. Shank, J. E. Rein, G. A. Huff and F. W. Dykes, "Facilities and Techniques for Analysis on Highly Radioactive Samples." *ibid.,* **29,** 1730-1739 (1957).

11 C. E. Lamb, "The High-Radiation Level Analytical Facility at Oak Ridge National Laboratory." *Talanta,* **6,** 20-27 (1960).

12 M. B. Hawkins, "Equipment for Handling, Storage and Transportation of Radioactive Materials." In H. Blatz, Ed., *Radiation Handbook,* First Edition. McGraw-Hill, New York (1959), pp. 18-2-18-23.

13 W. G. Ruehle, Jr., "Separating Transplutonium Elements from Irradiated Materials." *Nucleonics,* **12** (11), 84-85 (1964).

14 L. F. Miller and E. M. Kinderman, "Equipment for Remote Chemical Separations." *Ibid.,* **12** (11), 82-83 (1954).

15 D. C. Stewart and H. A. Elion, Eds., *Remote Analytical Chemistry, Progress in Nuclear Energy,* Series IX, Volume 10. Pergamon, Oxford (1970): (a) F. W. Dykes and J. E. Rein, "Ordinary Weighing and Density Determination," pp. 11-34; (b) W. L. Maddox, "Remote Pipetting," pp. 59-82; (c) P. F. Thomason, "The Application of Titrimetry to Remote Analysis," pp. 83-105; (d) W. D. Shults, "The Application of Electrochemical Techniques to Remote Analysis," pp. 307-357; (e) C. E. Lamb, "Sampling, Sample Dissolution, Evaporation and Combustion," pp. 107-142; (f) R. J. Morrow, "Ion Exchange and Solvent Extraction," pp. 181-224; (g) R. H. Gaddy, "Spectrographic Analyses in Containment Facilities," pp. 255-287; (h) F. W. Dykes, W. A. Ryder and G. V. Wheeler, "Instrumental Methods—Spectrophotometry, Atomic Absorption and Flame Emission Spectrometry," pp. 289-305.

16 Chapter 3, Ref. 59.

17 Torsion Balance Co., 125 Ellsworth St., Clifton, N.J. 07012.

18 M. F. Adams, "In-Cell Weighing System at the Hot Fuel Examination Facility (HFEF)." (Chap. 4, Ref 2x), pp. 140-144.

19 E. S. Fleischer, A. B. Snyder, T. C. Parsons and P. W. Howe, "Remote Pipetters for Use in Hot Cell Enclosures." USAEC Report UCRL 9661 (1964).

20 M. A. Wade, D. W. Ostby, F. W. Dykes and A. L. Olsen, "Vacuum Bottle System for Sampling Radioactive Solutions." (Chap. 4, Ref. 2z), pp. 382-387.

21 R. J. Jones, Ed., *Selected Measurement Methods for Plutonium and Uranium in the Nuclear Fuel Cycle,* USAEC Div. Tech. Info., Washington, D.C. (1963).

22 E. P. Horwitz, C. A. A. Bloomquist, K. A. Orlandini and D. Henderson, "The Separation of Milligram Quantities of Americium and Curium by Extraction Chromatography." *Radiochim. Acta.,* **8,** 127-132 (1968).

23 E. P. Horwitz, C. A. A. Bloomquist, J. A. Buzzell and H. W. Harvey," The Separation of Microgram Quantities of ^{252}Cf and ^{248}Cm by Extraction Chromatography in a High Level Cave." USAEC Report ANL 7546 (1969).

24 U. L. Upson and E. J. Wheelwright, "An In-Cell Recording Optical Spectrometer." USAEC Report BNL 337 (1967).

25 Chapter 4, Ref. 2.

26 Chapter 4, Ref. 3.

27 O. J. Wick, Ed., *Plutonium Handbook, A Guide to the Technology.* Gordon and Breach, New York (1967). In two volumes.

28 G. N. Walton, Ed., *Glove Boxes and Shielded Cells*, Proceedings of the Symposium on Glove Box Design and Operation, Harwell. England, Feb., 1957. Academic Press, New York (1958).

29 E. Grison, W. B. H. Lord and R. D. Fowler, Eds., *Plutonium, 1960*, Proceedings of the Second International Conference on Plutonium Metallurgy, Grenoble, France, April, 1960. Cleaver-Hume Press, London (1961).

30 A. E. Kay and M. B. Waldron, Eds., *Plutonium, 1965*, Proceedings of the Third International Conference on Plutonium, London, 1965. Barnes and Noble, New York (1967).

31 W. N. Miner, Ed., Proceedings of the Fourth International Symposium on *Plutonium and other Actinides, Santa Fe, N.M., 1970*. Metallurgical Society and American Institute of Mechanical Metallurgical and Petroleum Engineers, New York (1970).

32 H. Blank *et al.*, Eds., Proceedings of the Fifth International Conference *on Plutonium and Other Actinides, Baden-Baden, Switzerland, 1975*. American Elsevier, New York (1976).

33 F. L. Cochran, A. S. Hilbert and P. T. Mattson, Jr., "Design Criteria for In-Cell Installation of Metallograph Stages." (Chap. 4, Ref. 2s), pp. 104–111.

34 R. Natesh, J. P. Bacca, S. B. Brak, D. S. Taylor and J. R. White, "Metallography Facility in the Hot Fuel Examination Facility." (Chap. 4, Ref. 2t), pp. 203–212.

35 E. L. Long, Jr. and J. L. Miller, Jr., "Installation of the Shielded Electron Microprobe Analyzer at Oak Ridge National Laboratory." (Chap. 4, Ref. 2s), pp. 132–141.

chapter 7

Radiation Effects

Radiation has been used for years as a probe for study of the structure of matter, but until the advent of nuclear reactors very little investigation had been made of the effects of radiation on the grosser physical properties of materials. Reactors not only supplied a radiation source of sufficient size for such studies but also made these very necessary in order to obtain data for rational enginering design of all the facilities and equipment needed in the nuclear fuel cycle. Intensive efforts were accordingly made through the late 1940s, the 1950s, and the early 1960s to obtain engineering data regarding the effects of radiation on hundreds of different materials, largely by empirical experiment. This general survey type of activity has decreased somewhat as the needed information has been obtained and current programs in the area now tend to be more specialized. The development of new reactor fuels requires extensive irradiation and testing projects, and NASA and DOD have strong interest in radiation effects under conditions peculiar to space. The high-energy physicists must also initiate new equipment component testing programs each time a more powerful accelerator is planned. Industrial process interests are varied, but perhaps strongest in the area of modifying organic polymer behavior. The use of radiation as a tool for examining the fundamental nature of matter has steadily become more important.

Rather curiously, there does not seem to be any single compilation of information on radiation effects on matter—perhaps because the quantity of available data is somewhat overwhelming. The Radiation Effects Information Center (REIC) at Battelle-Columbus for some years published a monthly "accession list" in which relevant publications were reviewed and abstracted. REIC also published a numberof useful reviews,[1] each dealing with a particular class of materials, as well as data summaries and state-of-the-art surveys. Unfortunately the Center stopped operations in the mid-1960s and the reports are now hard to locate. A very extensive program of materials testing was carried out in connection with the NERVA (nuclear rocket) project at about the same period. The results were classified at the time of original issue and not declassified until the early 1970s, appearing as TID/SNA[2] or AGC[3] reports. These documents are also not found on every library shelf. A series of volumes in journal format dealing with the fundamental mechanisms involved in radiation effects is sponsored by the Society of Metallurgy of the American Institute of Mining, Metallurgical and Petroleum Engineers.[4] Among the older references is a special Nucleonics report[5] which tabulates many useful data. Several articles in the *Reactor Handbook*[6] are good sources of information and are cited individually below.

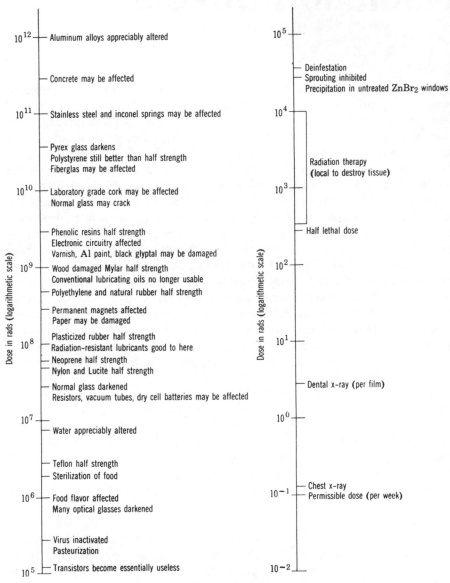

Figure 7.1 Approximate levels for radiation damage.

The approach in this chapter will be on the practical aspects of radiation damage from the point of view of the user scientists. A number of texts (only a few of which are cited) deal with the more basic aspects of the interaction of radiation with different types of matter.[7-11]

Figure 7.1 presents a generalized summary of selected radiation effects. Similar charts have been presented by Hennig[12] in terms of absorbed dose, and by Sisman and Wilson [5a] for neutron irradiations.

7.1 METALS AND ALLOYS

Changes in the physical properties of metals and alloys due to radiation will normally be non-observable by eye to the bench scientist and have little practical effect on his operations. Of more concern may be the activation of sample containers and experimental hardware, particularly by exposure to thermal neutrons. Aluminum cans are frequently used for holding samples for reactor irradiation, and at the time of removal may be at least briefly more radioactive by factors of 10 or more than the sample itself.

While irradiation effects on metals and alloys are of limited importance in the context of this book, changes of course occur and must be taken into consideration on the engineering level. The effects are largely due to 'knock-on' atoms that are displaced from their normal positions and become lodged elsewhere in the metallic lattice (interstitials), leaving vacancies at their previous sites. The results of these defects in a pure metal are similar to those produced by extensive cold working. The metal becomes harder and the ductility to brittleness temperature is raised. The elongation is reduced, but the tensile and yield strength normally increase. Thermal conductivity, density and expansion are usually not materially affected, but electrical conductivity may be decreased. These effects are less marked if the radiation exposure occurs at a higher temperature, and many of the changes can be largely reversed by post-irradiation annealing. Bowen and his associates[6a] present a clear general discussion of the mechanisms of changes induced by radiation in solids. These effects can actually be beneficial, depending on the application.

There are exceptions to the generalities given above, particularly for uranium metal. In this case the change in physical properties can be dramatic, probably due to local heating on the microscopic scale caused by the highly energetic fission products. Figure 7.2 shows the effect of equivalent amounts of reactor exposure on Th-U alloys containing varying amounts of uranium. The high-U alloy is essentially destroyed. Other alloying elements such as Mo, Zr, Si, etc., give some protection against such drastic effects. The prediction of the behavior of alloys in general is difficult and frequently can only be determined by experiment.

Shober[13] gives an extensive tabulation of experimentally observed radiation-induced changes in the tensile properties of a long list of metals and alloys. The same author has also produced an even more comprehensive survey as a REIC report.[1a]

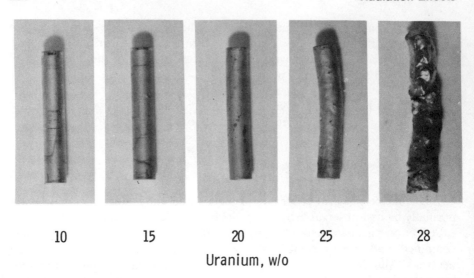

10   15   20   25   28

Uranium, w/o

Figure 7.2   Radiation effects on U-Th alloys containing different levels of fissionable ^{235}U. (ANL Photo 106-6257.)

7.2 INORGANIC SOLIDS

The behavior of irradiated glass was considered in connection with hot-cell viewing windows in Chapter 4 where the effects chiefly discussed were induced coloration and buildup of electrical charge. Color darkening of laboratory glassware can also be a problem on occasion, but is rarely serious unless the exposure has been at high levels for an extended period. A practical consideration in handling intensely active alpha emitters such as ^{242}Cm is however the destructive effect of the particule bombardment on glass surfaces—they are literally eaten away. A solution of an intense alpha emitter stored in a glass container will rapidly accumulate a silicate precipitate from SiO_2 molecules removed from the walls unless the nuclide is in very dilute form. The glass is of course weakened, and centrifugation of a tube after it has been in contact with such a solution for a few days will quite possibly result in breakage. Stock solutions of alpha emitters, if stored in glass at all, should be in dilute solution and the container carefully examined periodically for signs of trouble. Storage of ^{252}Cf solution in glass would presumably be an even worse situation since the container surfaces would be exposed to the even more destructive fission fragments as well as to alphas. Tuch[14] made a systematic study of the effect of alpha emitters on glass.

Glass is a noncrystalline solid. The behavior of irradiated crystalline inorganics has received extensive fundamental study, particularly for the alkali halides, quartz, semiconductor materials and graphite. The last of these (which in some ways can be considered as a metal) has received attention because of its importance as a reactor moderator, but its complex behavior under irradiation is even now not completely understood. Because of the two-dimensional character of the internal structure of graphite crystals the presence of interstitials causes an expansion in one direction. In a commercial polycrystalline graphite the crystals will be oriented in various ways, so the expansion is in all directions, a very practical problem for the reactor engineers.[6a]

The electrons surrounding the atoms of an ionic crystalline solid are tightly bound so such materials are generally good electrical insulators. Radiation on such solids will create free excited electrons and leave holes in their previous locations. Both of these effects will increase the electrical conductivity and the material will lose its insulating ability to some degree. This will be largely quickly restored when the radiation source is removed.[6a] Calkins and Schall[6b] indicate that covalently bonded ceramics are more affected by radiation than those of the ionic type. Typical observed changes are increased heat content, x-ray line broadening, and changes in thermal conductivity. These effects can often, but not in all cases, be annealed out by postirradiation treatment.

The action of radiation upon semi-conductors is very complex and continues to receive intensive study because of the importance of this type of material to modern technology. The effects can be broadly classified into two categories, those due to ionization (formation of ion pairs, hole-electron pairs, and free radicals) and those due to displacement of atoms, or in the case of fast neutrons, clusters of atoms. Neutron irradiation can also introduce an additional complication in that some of the impurity atoms critical to the operation of many semiconductors can be transmuted into new elements by the (n,γ) capture reaction. Changes due to this last mechanism will be permanent, other effects can often be removed by postirradiation annealing.[6a] A number of relevant texts[15-17] are available.

7.3 ELECTRONIC COMPONENTS

A recent issue of the IEEE (Institute of Electrical and Electronic Engineers) Proceedings[18] is devoted to effects of energetic radiations on electronic materials at the fundamental mechanisms level. As would be expected, several of the papers deal specifically with effects in semiconductors.

The behavior of electronic components in high-radiation fields is of critical importance to designers and users of high-energy accelerators. One of the largest of these is at CERN in Switzerland. Their Radiation Group has accordingly made a systematic study[19] of radiation-damage effects in various types of electronic equipment. Some of their tests were conducted by exposures in the swimming pool reactor at Seibersdorf in Austria, others by placing the samples adjacent to one of the beam dumps of the CERN 400-GeV Super Proton Synchrotron. Most of the irradiated materials were from European manufacturers. As is usual in such a case, different samples of similar items exhibited a range of stability behavior, partially accounting for the overlaps in the bar graphs shown in Figure 7.3, taken from the CERN report. The authors make the following comment: 'The results. . .lead to the overall conclusion that the

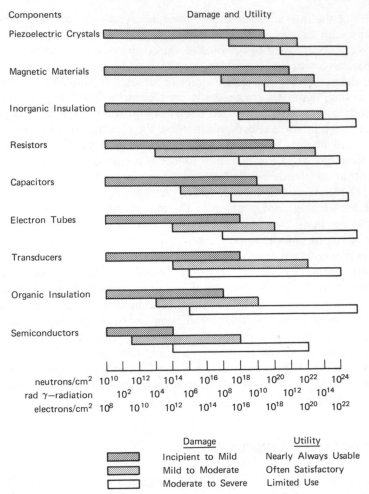

Figure 7.3   Radiation damage to electronic materials and components by fast reactor neutron, ^{60}Co gamma and 3-MeV electron radiation. (From Ref. 19.)

operation of electronic components and circuits is seriously affected by radiation environments with doses in the order of 10^{13} n/cm^2 or 10^4 rad (RPL); some components and circuits fail completely at doses of $10^{14}n/cm^2$ or 10^5 rad (RPL).' (The RPL indicates that the measurements were by radiophotoluminescent dosimeters.)

The construction material used for a particular type of component had considerable influence in some cases in the reactor irradiations. Metal film and potmetal cermet resistors, for example, were still stable after 3×10^6-rad exposure, whereas carbon and wire-wound types started to show damage effects at 3×10^5. Mica, MKL, and tantalum capacitors were good to the higher level; polyester was affected at 3×10^5; and ceramic, polycarbonate and Al electrolyte types at still another factor of 10 lower. Three of the diodes (Si, hot carrier;

Si, Zenel; and Ge, tunnel) went to 3×10^5. Other diodes and all of the other reactor-tested components were stable only to 3×10^4 rads or lower. (The irradiation conditions of course involved both gammas and neutrons.) Integrated circuits such as amplifiers, oscillators, power supplies, etc., showed instabilities earlier than individual components almost uniformly.

Older reviews dealing with radiation damage to electronics on the whole appear to report somewhat more optimistic results than the CERN group. Such references include a Livermore handbook prepared by Laine,[20] a REIC report,[16] and a paper by Shelton and Kenney[5b] in the previously cited Nucleonics Special Report. Calkins and Schall[6b] also report on some tests on electronic items.

7.4 OILS AND LUBRICANTS

The irradiation of organic compounds can cause polymerization, depolymerization, cross linking, gas emission, and/or molecular degradation owing to bond breakage or free-radical reactions. The effects on a particular organic species will depend primarily on its molecular configuration, but may also be affected by the temperature of the irradiation, the type and rate of dose delivery of the radiation, the presence or absence of oxygen, and the nature and concentration of any other molecules present.

Calkins and Schall[6b] present a tentative list of base lubricating materials in order of increasing radiation stability as: silicones, esters, mineral oils, ethers and alkyl aromatics. They also present a table taken from a classified document (NAA-SR-1304) by Durand and Faris, shown here in modified form as Table 7.1. Irradiated oils generally show marked increases in viscosity while greases are softened. Viscosity is thus the chief criterion used in judging damage to such materials. The threshold values given in the table are at the point where viscosity changes were first observed and the limiting values are at the point where the material solidified. (The neutron flux used was at thermal energies, the photons had an average energy of about 0.75 MeV.)

Exposure to air during irradiation accelerates oxidation and would reduce the values shown in Table 7.1 by a value of approximately 2. Oxidation inhibitors help in this situation. Other additive types that have been studied are radical inhibitors, protective compounds and thickening agents. Organic compounds containing sulfur or selenium are good antioxidants, and organic iodine species such as iodobenzene act as radical scavengers and can double or even triple the useful life of some of the base oils. The thickening agents decrease in viscosity when irradiated, so a judicious addition to a base material can compensate for the expected increase in radiation-induced viscosity in the oil itself.

Conventional metallic soap–mineral-oil greases soften markedly at about 10^{18} photons/cm². Lifetime increase of perhaps a factor of 10 can be obtained by special formulations, the recipes for three of which are given by Calkins and Schall. There does not appear to have been much study of solid lubricant

Table 7.1   Estimated Radiation Tolerances

| Compound Class | Dosage (per $cm^2 \times 10^{-18}$) | | | |
| | Threshold | | Limiting | |
	Neutrons	Photons	Neutrons	Photons
Polynuclear aromatics	1.8	12	15	90
Short-chain alkyl aromatics	1.5	2.5	7	30
Long-chain alkyl aromatics	0.5	2.0	2.5	10
Aliphatic polyethers	0.3	1.5	1.0	7.5
Aliphatic hydrocarbons	0.2	1.3	0.8	5
Aliphatic diesters	0.2	0.5	0.8	5
Aliphatic silicones	<0.1	<0.3	<0.5	<1

Source: Reference 6b.

systems such as Teflon particles or molybdenum disulphide suspended in a carrier matrix. Teflon is affected by radiation at relatively low doses so a lubricant using it as a base would be expected to show early changes in behavior.[6b]

REIC Report No. 19[1c] deals with the effect of radiation on lubricants.

7.5   PLASTICS AND ELASTOMERS

The need for gloves, booties, and other personnel protection devices; beta-shields; glovebox windows; containment-box fabrication materials; anticontamination shields; bag-outs; and a number of other specialized items means that the use of plastics and elastomers is even more prevalent in radioisotope work than in most other types of laboratory operations. The behavior of such materials when irradiated accordingly received early attention. Elastomers and plastics chemically are polymers in which repeating units of the same group of atoms ($-CH_2-CH_2-$ in polyethylene as an example) are joined together in long chains. In general, the effect of radiation is either to increase bridging (cross linking) between the chains or to cleave the chains into smaller units. The observable effects in these two cases will be very different—cross linking causes hardening and increased brittleness, while cleavage effects are almost the opposite. Other changes may also occur such as the grafting of side units onto a chain, molecular rearrangement, or complete decomposition.[11]

The following discussion is primarily based upon REIC 21[1d] which gives much more information than can be quoted here. Figure 7.4 is a composite of selected data taken from three of the bar graphs in that report. The exposures were to gamma irradiation. The open portion of a bar is the dosage range over which the damage is incipient to mild. The crosshatched interval indicates mild to moderate damage, with the material possibly still usable for some applications. The filled-in portions of the bars give the range in which the damage is severe. The ranges shown must be used as approximations and will not always be in point-by-point agreement with other reports owing to the different exposure conditions used by different investigators.

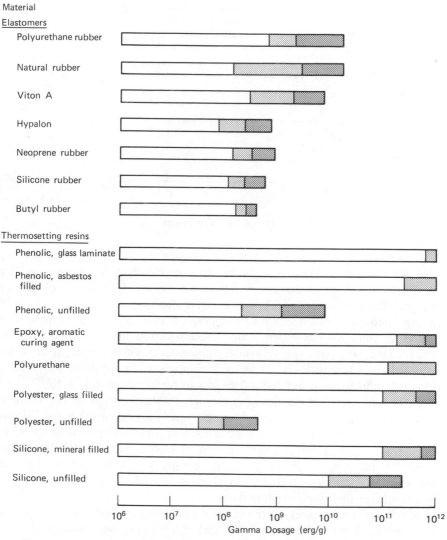

Figure 7.4   Radiation damage to elastomers and plastics. (Abstracted from Ref. 1d.)

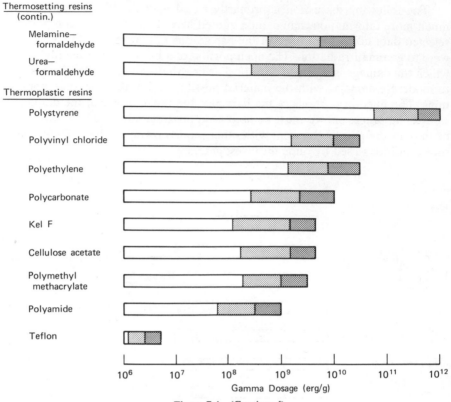

Figure 7.4 (Continued)

It is evident from Figure 7.4 that as a class the plastics, particularly the filled thermosetting types, tend to be more stable than the elastomers. The improvement gained in the thermosetting plastics by the addition of an inert filler material is pronounced. Teflon (in air) is affected very quickly at about 10^4 accumulated rads and decomposes to give off corrosive F_2-HF fumes. PVC (polyvinyl chloride) has much better resistance, but upon decomposing will produce HCl gas. Saran (polyvinylidene chloride) is not shown in the table, but is quoted elsewhere in REIC 21 as having a damage threshold at about 4.1×10^8 ergs/g and to have suffered a 25% loss in properties at 4.5×10^9.

Parts constructed of plastics or elastomers frequently show better resistance if their use is such that they can be immersed in an oil or other fluid during irradiation. Teflon insulation will retain useful properties up to about 5×10^9 ergs/g under such conditions and Viton A or silicone seals to 10^{10}. Irradiation at 500 °F has little effect on phenolic laminates as compared to the same treatment at room temperature, but there is a modest loss in resistance for epoxy, silicone and polyester laminates.

Much of the early work on the radiation behavior of elastomers and plastics was done by Bopp and Sisman,[21,22] at Oak Ridge. They derived an interesting ranking of the stability of the repeating units in the polymeric chains on the basis of their experience. The basic units of the two most stable polymers, polystyrene and aniline formaldehyde, both have benzene side groups which appear to act to give added stability. When the ring is in the main chain itself, as in phenol formaldehyde polymer, its presence apparently increases cleavage since unfilled phenolics tend to become crumbly upon radiolysis. The sugarlike ring unit of cellulose shows the same effect even more strongly. The $-CH=CH-$ unit present in many elastomers does not appear to behave much differently than the $-CH_2-CH_2-$ of polyethylene, indicating that the degree of chain unsaturation does not seem to be a strong influencing factor.

The early paper by Harrington[5c] in the Nucleonics Special Report tabulates a very considerable amount of information regarding radiation effects on elastomers and plastics. The article immediately following by Bresee and his associates[5d] dealing with coatings and gaskets also contains pertinent data.

A caution was previously expressed concerning the storage of very active solutions, particularly alpha emitters, in glass for extended periods. The same warning should be given for plastic containers. Here the problem is not so much the removal of surface material as it is embrittlement of the plastic, leading to eventual cracks and leakage. A stainless-steel beaker acting as a secondary container is highly advisable if either glass or plastic must be used for storage, even for relatively short periods.

7.6 ORGANIC INSULATORS

Electrical insulation on wiring is a special application of many plastics and elastomers, although the most radiation-resistant types are inorganic-organic formulations such as those having a glass or mica base, varnished or pasted over with silicone. Insulation is exposed to extremes of radiation and heat when wiring must be led to instrumentation in a thermal reactor. The exposure conditions will be even more severe in breeder and CTR reactors. The situation is more extreme if the wiring must be flexed while in use since materials under mechanical stress almost invariably show decreased radiation resistance as compared to static exposure.

The Radiation Group at CERN has made tests[23] on the radiation and fire resistance of cable insulating materials and is working closely with European manufacturers to develop improved products. Loading plastics or elastomers with small amounts of a filler such as carbon black, asbestos, mica, clay, kaolin, chalk, zinc oxide, silica, etc., increases stability. Antirads are also of benefit, but the one chosen must be compatible with the base polymer. Their recommendations on antirads are reproduced in Table 7.2. The authors concluded that radiation did not appear to have much influence on the fire resistance of the materials they tested.

The Japanese have carried out an extensive program of insulation testing in connection with a joint Japanese-American fast-breeder-reactor development

Table 7.2   Antirads for Insulator Materials

Polymer	Antirad
Styrene butadiene rubber	Alpha naphthylamine
Nitrile rubber	Alpha naphthylamine
	Quinhydrone
Hypalon	Quinhydrone
Polyvinylchloride	Phthalic acid ester
Polyethylene	Amines
	Sulfur compounds
	Phenolic compounds

Source: CERN 75−3, Ref. 23.

program. Their report[24] is available in translation and the results appear to be in general agreement with those that would be predicted from Figure 7.4. As usual, however, supposedly similar materials from different suppliers (or even different samples from the same supplier) show a range in stability. In other words, the manufacturing process (monomer source, plasticizers used, curing techniques) has to be added to the many other variables affecting the resistance to radiation damage of a particular plastic or elastomer.

Calkins[25] has stated that a threshold value of 4.5-7 $\times$ 10[6] rads exists for radiation damage to organic electrical insulators, and Calkins and Schall[6b] have tabulated the stability results for such materials as reported by a number of investigators. The authors comment that fiberglass impregnated with aromatic silicones has good resistance, although some gas is evolved during irradiation. The phenolics appear to be satisfactory for many applications.

Prise[26] has discussed electrical insulating materials and points out that in normal practice a major consideration in selecting an insulator is often its resistance to high temperatures. Some materials having good radiation stability may have to be eliminated if they are to be used in high-temperature environments. He cites natural rubber as an example of such a situation. He emphasizes improvements that can be gained by adding fillers and antirads (antioxidants) to process formulations. Teflon irradiated in vacuum shows an appreciably higher stability than when given the same treatment in air.

Calkins and Schall present some data on the radiation stability of thermal-insulation samples. There was some gas evolution, but little other evidence of damage.

(Two very comprehensive reports pertaining to this section and the one immediately previous were just received from the Radiation Group of the Health and Safety Division at CERN in Switzerland. The first[27a] of these documents deals with cable insulation materials, the second[27b] with thermosetting and thermoplastic resins. Each of the reports summarizes an extensive testing program, primarily on samples submitted by European manufacturers. Very useful data tables giving the physical properties of pertinent unirradiated materials and notations on their basic chemical compositions are included. Time has not allowed a thorough analysis of these reports, but it would appear that on the whole the authors to a modest degree would predict some extension of the useful lives of at least some materials beyond the limits shown in Figure 7.4. The test data for each individual item are shown in the reports in graphs of dose received versus changes in critical physical properties.)

7.7 CHEMICAL OPERATIONS

Chemical reagents and materials such as ion-exchange resins do not always behave normally when in the presence of radiation. As would be expected from the relative LET values, alpha emitters generate more of these types of problems in the comparatively small volume equipment characteristic of research than do betas, some of which escape but still create more changes than do gammas which will largely pass entirely out of the apparatus before losing much energy. An extreme example of trying to do chemistry with an intense alpha emitter is with ^{242}Cm whose solutions have been likened to freshly opened carbonated cola.

Zittel[28] has reviewed the effects of radiation on common analytical reagents and has tabulated the relative susceptibility to radiation damage of some of the more frequently used chemicals. Sulfuric, phosphoric and acetic are rated as being of low susceptibility among the acids; perchloric, medium; hydrochloric and nitric, medium to high; and formic and oxalic, high. Similar ratings for commonly used oxidants are: low to medium susceptibility, Ce(IV), Cr(VI), V(V), Co(III) and Tl(III); the permanganate ion, medium to high. For reductants the rankings are: low susceptibility, bromide ion and U(VI); low to medium, iodide ion; medium, sulfite, phosphite and ferrocyanide; medium to high, hydroxylamine, hydrazine, nitrite, and arsenite; and high, Sn(II). Organic complexing agents and indicators generally fall in the medium to high classification, but with EDTA and Arsenazo III being rated as low to medium in radiation susceptibility.

The most important chemical reagent of all is water. As has been noted earlier, the primary ionization event in irradiated water is thought to be

$$2H_2O \rightarrow H_3O^+ + e_{aq}^- + OH.$$

The hydrated electron thus produced is an extremely powerful reducing agent, whereas the hydroxyl radical is a comparably potent oxidant. The hydrated electron has a very short lifetime (longer in neutral or alkaline solutions than in acid), but produces atomic H, another strong reductant, when the decay occurs in aqueous solution. A whole series of additional reactions may follow these first events to yield H_2, O_2, H_2O_2, HO_2, etc. The possibilities of reaction between these active species and any solute present are obvious.

The radiation chemists state yields of their reactions in terms of 'G values,' defined as the number of molecules produced or decomposed per 100 eV of energy absorbed. Using the relationship

$$G \text{ value} \times 1.01 \times 10^{-9} \text{ (moles/l)/R} = \text{concentration change,}$$

allows the estimation of chemical changes in terms more familiar to most other chemists. G values for numerous reactions have now been determined and many are tabulated in the various general references on radiation chemistry previously cited and Allen[29] has prepared an extensive compilation. Such data must however be used with some care. G values are normally determined in simple systems, perhaps containing only the chemical species of interest. The values can be markedly changed by variations in pH, solute concentrations, the presence of other solutes or air, temperature, total dose and its delivery rate, etc.

Radiation effects on organic solvents have not been as thoroughly studied as have aqueous systems, although the use of pulse radiolysis techniques[30] is allowing the constant addition of new data in this area. The free-hydrogen atoms produced by radiolysis may abstract hydrogen from some of the solvent molecules, or may add onto aromatic structures. Polymerization effects may lead to emulsion formation in two-phase water-solvent systems. Free-radical reactions evidently play a considerable part in solvent radiolysis, but detailed mechanisms are still uncertain in most cases. Even less is known about the behavior of solutes in such solvents. A dilute TTA (thenoyltrifluoroacetone) solution in xylene exposed to 2×10^6R of gamma irradiation showed little change, but there was evidence of damage to the side chain of the xylene.[31] Tributyl phosphate (TBP) in hydrocarbon solvents has been used extensively in reactor-fuel reprocessing as a means of separating uranium, plutonium and thorium from fission products. A primary effect of radiation on TBP is thought to be hydrolysis to di- and mono-butyl acid forms. The presence of these hydrolytic products, even in very small amounts, markedly decreases the decontamination from the fission products and increases the difficulty of stripping the plutonium back out of the solvent. The G value for the hydrolysis reaction has been estimated as being about 0.2. At about 10 W·hr/l emulsification difficulties make the separation processes inoperable.[31]

Pure hydrochloric acid solutions less than $1M$ in concentration are not markedly affected by gamma irradiation, but as the molarity is raised, acid depletion begins and increases to the point where the G value for the loss of hydrogen ion approaches 1. Atomic Cl is thought to be involved since the effect is not seen in most other acids and the presence of added chloride salts increases the acid depletion rate. Added methyl alcohol (1%) or chloral hydrate act to protect the acid loss to some degree. The effect of irradiation upon nitric acid is first to reduce the nitrate ion to nitrite. Sulfuric acid in less than 60% concentration is relatively stable, but some decomposition occurs above that level. Phosphoric acid is exceptionally stable towards radiation. Perchloric acid is decomposed into chlorate and oxygen. Formic acid is very susceptible to radiolysis as is oxalic acid. Acetic acid is apparently more resistent.

EDTA finds use both in analytical chemistry and in some separations in radiochemistry as a complexing agent. The compound is relatively stable towards moderate levels of radiation if in dilute solution, but decomposes at higher exposures. Cathers[31] gives a G value of about 1.7 for this process. Zittle[28] estimates a figure of 1.4 for solutions in the $0.005M$-$0.01M$ range. A solution at the first concentration will be about 15% degraded by exposure to 5×10^5 R of gamma radiation.

Solutes that react preferentially with free radicals can sometimes be added to solutions to act as protective agents for other susceptible species present. These 'scavengers' are frequently used in radiation chemistry when it is desired to suppress the action of one type of free radical so that the effect of others can be more readily observed. (N_2O converts e_{aq}^- to OH, for example, or alcohol can be used to convert OH to an alcohol radical, leaving e_{aq}^- or H predominant in the solution.[32]) Aromatic compounds mixed in with other organics can act as protective agents in some cases. Their resonating structures apparently permit absorption of some of the energy originally imparted to the molecule being protected.

Ion-exchange resins are cross-linked organic polymers having reactive sites that retain either cations or anions, depending on their formulation. Radiation can attack the cross-linkage structure, resulting in column-bed swelling or even charring when very high alpha concentrations are used. The reactive sites can also be affected, resulting in a loss of absorptive capacity. Another practical problem is the formation of gas bubbles in the eluting solution, leading to channeling in the bed, floating resin at the top of the column or blockage of the elutriant flow. Bubbling can be reduced to some degree by keeping the bed under pressure.

The quaternary amine anion resins are more susceptible to radiation damage than the strong acid cation types. The combination of anion resin in the nitrate form plus high alpha levels is particularly unstable and has resulted in explosive incidents.[33] Morrow[34] recommends that anion resins never be used in high-radiation fields with nitric acid concentrations higher than $7M$. Molen[35] adds the further recommendations that nitrated columns never be allowed to dry out and that the resin during disposal be treated as a potential explosive if it is in the dry state. Chloride form anion resins are more stable. Lopez-Menchero[36] found than 10^9 rads from a ^{60}Co gamma source reduced the capacity of an acidic sulphonate cation resin by over 50% and completely destroyed a strong base anion resin in the OH^- form. Nater,[37] in a very similar experiment, found a reduction in capacity of 20% in a cation resin, 50% in an anion resin. He concluded that the chief action for the former was bond-breakage, but that additional linkages were formed in the anion resin. Miles[33] cites a number of investigations of resin behavior under irradiation.

Shults[38] discusses the effect of radiation on electrochemical analytical methods at some length. There are two major sources producing errors, changes caused by the radiation in the solution itself and changes induced in the electrodes. Shults notes that conductance methods are more reliable than direct potentiometry for long-time monitoring of radioactive solutions. Voltammetric and polarographic methods have been successfully utilized for hot-cell analyses, as have coulometric techniques. Controlled potential coulometry is possibly the most widely used electrochemical analytical method for nuclear work.

A radiation-damage problem that is sometimes overlooked is the self-destruction of labeled compounds during storage. A molecule containing a radioactive label such as ^{14}C may not only destroy itself but also initiate chain reactions affecting many unlabeled neighbors. The text by Raaen and his associates[39] presents a good review of the problem with suggestions as to how it might be mitigated. An early survey of the self-radiation problem was made by Tolbert[40] who first summarized G values for different types of organic compounds irradiated *in vacuo*. These destruction coefficients ranged from 1 for aromatics to as high as 2000 for certain unsaturated hydrocarbons. Tolbert then calculated the self-imposed dose for compounds labeled with varying levels of active nuclides, finding, for example, that a ^{14}C-labeled material at the 100-mCi/g level would receive 7×10^6 rads dosage in a month, and a similarly labeled ^3H compound, 9×10^5 rads. He then calculated the percentage decomposition that would be expected in a hypothetical compound of 125 molecular weight as a function of the dosage at different G values. The decomposition began to be appreciable for integrated exposures above 10^7 rads and destruction G values greater than 5.

REFERENCES

1 Radiation Effects Information Center, Battelle Memorial Institute, Columbus, Ohio 43200: (a) F. R. Shober, 'The Effect of Nuclear Radiation on Structural Metals.' REIC Report 20 (1961); (b) R. K. Thatcher, D. J. Hamman, W. E. Chapin, C. L. Hands and E. N. Wyler, 'The Effect of Nuclear Radiations on Electronic Components Including Semi-Conductors.' REIC Report 36 (1964); (c) S. L. Cosgrove and R. L. Dueltgen, 'The Effect of Nuclear Radiations on Lubricants and Hydraulic Fluids.' REIC Report 19 (1961); (d) R. W. King, N. J. Broadway and S. Palinchak, 'The Effect of Nuclear Radiations on Elastomeric and Plastic Components and Materials.' REIC Report 21 (1961); N. J. Broadway and S. Palinchak, (Same title), REIC Report 21 (Add.) (1964).

2 Anon, 'Radiation Effects: General.' USAEC Report TID/SNA 2256 (1963); declassified September, 1973.

3 Anon, 'Radiation Effects Data Book.' USAEC Report AGC 2277, Vol. 2 (Rev) (1963); declassified August, 1973.

4 W. F. Sheely, Ed., *Radiation Effects*, Gordon and Breach, New York. (A series in journal format. Volume 40 appeared in 1979).

5 'How Radiation Affects Materials, A Special Report,' *Nucleonics, 14* (9), 53-88 (1956): (a) O. Sisman and J. C. Wilson, 'Engineering Use of Damage Data,' pp. 58-62; (b) R. D. Shelton and J. G. Kenney, 'Damaging Effects on Electronics Components,' pp. 66-69; (c) R. Harrington, 'Damaging Effects of Radiation on Plastics and Elastomers,' pp. 70-74; (d) J. C. Bresee, J. R. Flanary, J. H. Goode, C. D. Watson and J. S. Watson, 'Damaging Effects of Radiation on Chemical Materials,' pp. 75-81.

6 (*Reactor Handbook*, Chap. 4, Ref. 3): (a) D. B. Bowen, C.E. Dixon and J. E. Hove, 'Generalities of Radiation Damage,' Vol. I, pp. 40-52; (b) G. D. Calkins and P. Schall, 'Radiation Damage-Miscellaneous Materials,' Vol. I, pp. 74-83.

7 A. J. Swallow, *Radiation Chemistry, An Introduction*, Wiley, New York (1973).

8 J. F.Kirchner and R. Bowman, *Effects of Radiation on Materials and Components*, Reinhold, New York (1964).

9 J. G. Carrol and R. O. Bolt, Eds., *Radiation Effects on Organic Materials*, Academic Press, New York (1971).

10 I. G. Draganic and Z. D. Draganic, *The Radiation Chemistry of Water*, Academic Press, New York (1971).

11 J. E. Wilson, *Radiation Chemistry of Monomers, Polymers and Plastics*, Marcel Dekker, New York (1974).

12 G. H. Hennig, 'Moderators, Shielding and Auxilliary Equipment.' In J. J. Harwood *et al.*, Eds., *The Effect of Radiation on Materials*, Reinhold, New York(1958).

13 F. R. Shober, 'Reactor-Radiation Effects on Structural Materials.' *Nucleonics*, 20 (8), 134-140 (1962).

14 D. G. Tuck, 'Radiation Damage to Glass Surfaces by α-Particle Bombardment.' *Int. J. Appl. Radia. Isotopes*, 15, 49-57 (Feb. 1964).

15 F. Larin, *Radiation Effects in Semiconductor Devices*, Wiley, New York (1968).

16 J. W. Corbett, *Electron Radiation Damage in Semi-Conductors and Metals*, Academic Press, New York (1966).

17 R. J. Chaffin, *Microwave Semiconductor Devices. Fundamentals and Radiation Damage*, Wiley, New York (1973).

18 'Special Issue on the Effects and Uses of Energetic Radiations on Electronic Materials.' *Proc. IEEE* 62 (9), 1187-1277 (1974).

19 S. Battisti, R. Bossart, H. Schönbacher and M. Van de Voorde, 'Radiation Damage to Electronic Components.' Report CERN 75-18 (1975).

20 E. F. Laine, 'Radiation Effects on Electronic Components.' USAEC Report UCID 4544 (1962).

21 C. D. Bopp and O. Sisman, 'Radiation Stability of Plastics and Elastomers (Supplement to ORNL 928).' USAEC Report ORNL 1373 (1953).

22 C. D. Bopp and O. Sisman, 'Radiation Stability of Plastics and Elastomers.' *Nucleonics*, 13 (7), 28-31 (1955).

23 H. Schönbacher and M. H. Van de Voorde, 'Radiation and Fire Resistance of Cable-Insulating Materials Used in Accelerator Engineering.' Report CERN 75-3 (1975).

24 J. Kakuta, N. Wayakama and T. Kawakame, 'Study of Fast Breeder Instrumentation. II. Gamma Irradiation Tests of Wire and Cable Insulations and Electronic Component Materials.' JAERI Report JAPFNR-172 (1974).

25 V. P. Calkins, 'Radiation Damage to Nonmetallic Materials.' USAEC Report APEX 172 (1954).

26 W. J. Prise, 'When the Gamma Heat Is on Insulators.' *Elec. Design,*, 11, 72-75 (May 23, 1968).

27 H. Schönbacher and A. Stolarz-Izycha, 'Compilation of Radiation Damage Test Data': (a) 'Part I: Cable Insulating Materials.' Report CERN 79-04 (1979); (b) Part II: Thermosetting and Thermoplastic Resins.' Report CERN 79-08 (1979).

28 H. E. Zittel, 'Effect of Radiation on Common Analytical Reagents.' (Chapter 6, Ref. 15, pp. 395-412.)

29 A. O. Allen, 'Yields of Free Ions Formed in Liquids by Radiation.' USNBS Report PB 255004 (1976).

30 Chapter 2, Ref. 23.

31 G. I. Cathers, 'Radiation Damage to Radiochemical Processing Agents.' *Proceedings of the International Conference on Peaceful Uses of Atomic Energy, Geneva, 1955*, 7, 490-495 (1956), A/CONF 8/P743, United Nations, New York.

32 Drs. Sheffield Gordon and Myran Sauer, Argonne National Laboratory, Personal communications.

33 F. W. Miles, 'Ion-Exchange-Resin System Failures in Processing Actinides.' *Nucl. Safety*, 9 (5), 394-406 (1968); *Isot. Rad. Tech.*, 6 (4), 428-440 (1969).

34 Chapter 6, Ref. 15f.

35 G.F. Molen, 'Compatability Studies of Anion Exchange Resins. I. The Compatability of Dowex 1×4 Nitrate Form Anion Resin with Nitric Acid and Oxides of Nitrogen.' USAEC Report RFP 531 (1965).

36 E. Lopez-Menchero, 'Recent Research and Development Work in ENEA Countries Related to the Treatment of Low and Intermediate Level Waste.' USAEC Report NP 15946 (1965).

37 K. A. Nater, 'Radiation Damage in Ion Exchangers.' *Atoomenergie Haar Toepassingen*, 4, 155-163 (1962).

38 Chapter 6, Ref. 15d.

39 Chapter 3, Ref. 56.

40 B. M. Tolbert, 'Self-Destruction in Radioactive Compounds.' *Nucleonics*, 18 (8), 74-75 (1960).

chapter 8

Nuclear Criticality

Some of the tables and graphs presented in this chapter were compiled or taken from different sources so small internal discrepancies may be noted. The accepted numbers have changed over the years (generally in a minor way) as better computational methods and experimental data have become available. Unless noted, no safety factors have been applied to any of the numbers given, and in some cases the various qualifiers (there are always at least several) have not been completely stated. The data presented are quoted for illustrative puposes so the chapter has only limited value as a practical reference manual in real-life situations.

8.1 THE FISSION PROCESS

As one goes up the Periodic Table to the heavier chemical elements (for practical purposes, thorium and above), the nuclei become more fissionable, that is, they can be made to break apart into two fragments (the fission products) of more or less comparable size. This breaking-up process is generally brought about by absorption of a neutron from the outside into the atom's nucleus [the (n, f) reaction]. Fission can also be made to occur by bombardment of the nucleus with charged particles of sufficiently high energy as in a cyclotron but such reactions are not involved in the criticality problem as discussed here. Very heavy nuclei will also occasionally fission spontaneously. This last process depends on the internal structure of a particular nucleus so isotopes, even of the same element, can have widely differing spontaneous-fission half-lives. These can vary from billions of years for an isotope such as ^{240}Pu down to a few hours for a very heavy nuclide such as ^{256}Fm.

Fission does not always occur when a neutron is absorbed into the nucleus of a fissile atom but is only one of several possible nuclear reactions that may take place. By far the most important of these competitive reactions is neutron capture, the (n,γ) reaction. This reaction may also occur with the nuclei of any other atoms that happen to be in a reactor fuel, processing solution, etc., in association with the fissionable material. Such captured neutrons are of course lost in terms of producing further fission.

Whether or not a heavy-element nucleus can be fissioned by both low-energy (slow or thermal) neutrons and high-energy (fast) neutrons—the *fissile* isotopes—or by fast neutrons only—the *fissionable* isotopes—depends on the internal structure of the nucleus. This structure also determines the probability that a particular reaction will occur, that is, fission in preference to capture or vice versa. These probabilities are stated in terms of cross sections, σ_f for fission and σ_c for capture. The values of these cross sections will vary markedly, depending on the energy of the incoming neutron. Fissile atoms have very high fission cross sections for slow neutrons, but σ_f drops very rapidly with increase in neutron energy. Fissionable atoms on the other hand are not fissioned by thermal neutrons to any extent, but have appreciable cross sections for fast neutrons. The result of this pattern is that all of the heavy-element isotopes have a relatively low but relatively uniform tendency to fission with fast neutrons, but dramatic differences occur if the neutron flux has been thermalized by scattering processes. Thus the fissile isotopes—the ones useful as thermal reactor fuels such as ^{235}U, ^{239}Pu, and ^{233}U—have large fission cross sections for slow neutrons, whereas the comparable values for ^{238}U and ^{240}Pu are essentially zero. On the other hand all of these nuclides will have fission cross sections in the several barns range in a fast breeder reactor.

There are of course other heavy-isotope fissiles besides those named above, but at present only four are available in sufficient quantity to create a criticality problem. This situation will not necessarily be true in the future so critical masses for other isotopes are included in Tables 8.4 and 8.5 which will be discussed later. The present major concern will be with the big four: ^{233}U, ^{235}U, ^{239}Pu, and ^{241}Pu. [Pu-241 is available in pure form in only very small quantities, but builds up in a high burn-up (highly irradiated) fuel and its presence must be taken into account.]

Several important things happen when a nucleus does fission, whether by neutron absorption or spontaneously. A certain amount of mass is converted to energy which can be used constructively as in a power reactor or destructively as in a nuclear bomb. Some free neutrons are also released during the process in addition to the primary fission products. As an example, about 2.5 neutrons are emitted on the average when ^{235}U fissions, the comparable figure for ^{239}Pu being approximately 3. Using the ^{235}U example, one neutron goes in and 2.5 ('first-generation') neutrons come out for a net gain of 1.5 per fissioned atom. Some of the emitted neutrons will be captured through the (n,γ) reaction by the ^{238}U generally associated with the ^{235}U, some by the ^{235}U itself and others by atoms of other elements present, and some will leak out of the material before they can react at all. If matters can be arranged to restrict losses so that, on the average, one of the first-generation neutrons causes fission in a new ^{235}U nucleus to produce a second generation of neutrons, and so on, a chain reaction will be established. If there is exactly one neutron left over per fission to initiate a new event, the system is in a steady state as in a reactor. If on the average there is less than one neutron, the chain reaction fizzles out and the material remains subcritical. On the other hand if there is more than one neutron on the average to cause new fissions, the chain reaction is a runaway and out of control and the material is in supercritical condition. Guaranteeing that this last does not happen by accident is the 'criticality problem' since tremendous amounts of energy, fission products and neutrons are almost instantaneously released.

(A nuclear bomb is specifically designed to obtain the maximum amount of energy release out of the supercritical situation before the chain reaction is stopped by the material being blown apart. This is accomplished by bringing subunits of the critical mass together with great force and speed in a carefully designed geometry. A criticality accident occuring during routine reactor-fuel handling or processing would not cause anywhere near as widespread destruction as a bomb, but could have lethal effects. Any exposed individual immediately nearby would be lucky to survive and the cloud of released fission products could produce a widespread hazard if not immediately contained.)

The fact that ^{239}Pu produces three neutrons per fission as compared to 2.5 for ^{235}U would seem to indicate that the problem of neutron losses would be less for the first isotope. The extra neutron yield for ^{239}Pu is however largely cancelled out by that nuclide's higher cross section for simple absorption. In other words, σ_c/σ_f (a ratio known as α in nuclear engineering) is higher and substantially more neutrons are lost to the fuel in the plutonium case:

Isotope	σ_c (barns)	σ_f (barns)	σ_c/σ_f (α)
^{235}U	101	577	0.175
^{239}Pu	340	810	0.42

The original neutrons that initiate the chain reaction obviously must come from somewhere. Very large amounts of fuel are required in a natural or slightly enriched uranium reactor, so in spite of the fact that spontaneous fission of ^{238}U produces only 15 (n/sec)/kg, a sufficient flux for startup is available.[1] Very much less fuel and very little ^{238}U are present in reactors using highly enriched uranium so a neutron source must be furnished. This could be a spontaneously fissioning isotope such as ^{252}Cf—a potent neutron emitter—but more generally is an (α,n) source. Compact neutron sources can be fabricated by mixing ^{239}Pu, ^{241}Am, etc., with a light element such as beryllium. Since most of the heavy-element isotopes are alpha emitters and since light-element contaminants are ubiquitous, there will generally be a very low but possibly sufficient flux of neutrons in many situations to act as an initiator if other conditions are right for starting a chain reaction, particularly in handling the very heavy elements where a low flux from spontaneous fission will also be present.

8.2 THE CRITICAL MASS

A simple first step in considering critical mass is to visualize a series of metal spheres composed of a pure fissile isotope such as ^{239}Pu or ^{235}U. Since they are spheres, the ratio of the external surface to the volume, and thus to the mass, will decrease as the balls grow larger. The neutrons present in the smaller spheres therefore have a better chance to leak out and a chain reaction cannot occur. As the spheres grow larger, however, a point will eventually be reached where the leakage rate is not sufficient to keep the material in a subcritical condition. This is the point of 'critical radius' and of 'critical mass.'

The spheres in this description are 'bare,' that is, considered to be hanging in space with nothing around them that could reflect the escaping neutrons back into the metal. This of course is not a realistic situation since the sphere must be supported somehow and probably in some sort of an atmosphere. The computer codes used to calculate critical masses thus generally consider three situations—bare, partially reflected (sometimes called nominally reflected), and totally or fully reflected. The partially reflected case considers that the sphere is surrounded by 1 in. of water or its nuclear equivalent. A full reflector is water at least 3 in. thick or its nuclear equivalent.[2] Obviously, the more neutrons reflected back, the smaller the sphere and mass of fissile metal at the critical point, so the fully reflected calculations are the most conservative and the basis of most recommended quantity limitations.

Table 8.1 Criticality Data for ^{233}U (Single Units)

Parameter	Metal (Density, 18.44)		Water Solution	
	Fully Reflected	Bare	Fully Reflected	Bare[a]
Minimum critical mass (kg)	6.7	17.0	0.550	1.2
Infinite cylinder (cm)	4.6	8.2	11.5	19.0
Infinite slab (cm)	0.54	4.6	3.0	10.2
Volume (l) [b]	0.407	0.84	3.5	8.7
Minimum critical concentration (g/l)	—	—	10.8	—
Areal density (g/100 cm²)	—	—	35.0	47.4

Source: References 2–4.

[a] One sixteenth-in. stainless-steel reflector assumed for all ''bare''-solution data in Tables 8.1–8.5.

[b] Spherical volume.

Table 8.2 Criticality Data for ^{235}U (Single Units)

| | Metal (Density, 18.44) | | Water Solution | |
Parameter	Fully Reflected	Bare	Fully Reflected	Bare
Minimum critical mass (kg)	20.1	47.0	0.76	1.40
Infinite cylinder (cm)	7.3	11.4	13.9	21.6
Infinite slab (cm)	1.3	5.6	4.6	11.4
Volume (l) [a]	—	2.7	5.8	14.0
Minimum critical concentration (g/l)	—	—	11.5	—
Areal density (g/100 cm ^{2})	0.13	—	40.0	56.0

Source: References 2−4.

[a] Spherical volume.

Table 8.3 Criticality Data for ^{239}Pu (Single Units)

| | Metal (Density, 19.5) | | Water Solution | | |
| | | | (N:Pu>4) | | No NO$_3$ |
Parameter	Fully Reflected	Bare	Fully Reflected	Bare	Fully Reflected
Minimum critical mass (kg)	4.9	10.2	0.51	0.905	0.51
Diameter-infinite cylinder (cm)	4.4	6.1	15.7	23.2	12.5
Thickness-infinite slab (cm)	0.66	2.8	5.8	13.5	3.3
Spherical volume (l)	0.28	0.51	7.7	—	4.5
Minimum critical concentration (g/l)	—	—	7.3	—	7.8
Areal density (g/100cm²)	—	—	25.0	—	—

Source: References 2−4.

Tables 8.1-8.5 present criticality mass data for the more important heavy-metal isotopes. No safety factors have been applied to any of the figures shown and no values are quoted for the nominally reflected situation. The data in Table 8.3 for plutonium metal are for the alpha form (density 19.5). More of

Table 8.4   Criticality Data—Other Fissiles[a]

Isotope	Metal Spheres		Water Solution[b]	
	Critical Mass (kg)		Minimum Critical Mass (g)	Approximate Concentration
	Fully Reflected	Bare	Fully Reflected	g/l
$^{241}_{94}$Pu	$\approx 5.2^c$	—	260	32[d]
$^{242m}_{95}$Am	≈ 3.8	8.4	23	5
$^{243}_{96}$Cm	—	—	213	40
$^{245}_{96}$Cm	—	—	42	15
$^{247}_{96}$Cm	—	—	159	60
$^{249}_{96}$Cm	—	—	32	20
$^{251}_{98}$Cf	—	—	10	6

Source: References 5–7.

[a] $A - Z$ is odd.

[b] Spherical geometry.

[c] Density 19.5 (alpha-phase metal).

[d] H/X ratio ≈ 880. The limiting critical concentration for ^{241}Pu is 5.0 g/l.

the delta form (density 15.44) can be handled, that is, the critical mass for the latter, reflected, is 7.6 kg as compared to 4.9 kg for the alpha form. The details of these tables will be further discussed in later sections.

A solid metal sphere will have a lower critical mass than if the metal were as filings, powder, etc., simply because of the difference in overall density. The solid line in Figure 8.1 indicates the change in the critical mass of water reflected 93%-enriched uranium as a function of density. [The U(93) notation indicates the enrichment. The metal density is 18.75 in fully condensed form.] It will be seen that a 20% decrease in apparent density (due to its being as filings, for example) increases the critical mass from roughly 25 to 34 kg. Uranium compounds will have lower densities than the metal so will have larger critical masses, but the effect of the lower density is partially offset by the moderating

Table 8.5 Criticality Data—Fissionable Isotopes ($A - Z$ is Even)

| | | Metal Spheres | | | Water Solution | |
| | | Critical Mass (kg) | | | Limiting Critical Concentration (kg/l) | At Approximate H/X of: |
Isotope	Density (g/cm³)	Water Reflected	Steel Reflected	Bare		
$^{237}_{93}\text{Np}$	20.45	64.9	43.1	68.6	12.7	0.8
$^{238}_{94}\text{Pu}$	19.6	5.6	4.5	7.2	5.1	3.8
$^{240}_{94}\text{Pu}$	19.6	148.4	96.4	158.4	17.5	0.16
$^{241}_{95}\text{Am}$	11.7	105.3	71.6	113.5	7.6	1.5
$^{244}_{96}\text{Cm}$	13.5	22.0	14.2	23.2	6.7	2.0

Source: References 5 and 6.

effect of the other associated elements in the compound. The results of these opposing effects can be seen in Figure 8.1. The compounds always have higher critical masses than pure solid metal, but because of the moderating effect, never as high as, say, a sphere of uranium metal powder of comparable density.

8.3 OTHER PARAMETERS

Fissile materials in metallic form are not always so obliging as to be in the shape of spheres so calculations have to be made for other geometries. The surface-to-volume ratio of a cylinder will be different from that of a sphere, and if the diameter can be kept small enough so that sufficient leakage surface is available the cylinder can be of infinite length and mass and still be safe. The same is true of an infinite slab of the metal—if it is kept thin enough the surface-to-volume ratio will be such that the material must remain subcritical. Recommendations are accordingly usually made for these two simple geometrical shapes as well as for spheres; again bare, nominally reflected and fully reflected.

The situation where the fissile material is in solution is quite different in that the solvent (usually water) and any other constituents present, such as the acid used to dissolve the metal, will act as neutron moderators and rapidly reduce the average neutron energy down into the thermal range where the fission cross sections for the fissiles are much higher. It is for this reason that charts showing the change of critical mass with the concentration of the material are sometimes given in terms of the H/X ratio, effectively the ratio between the number

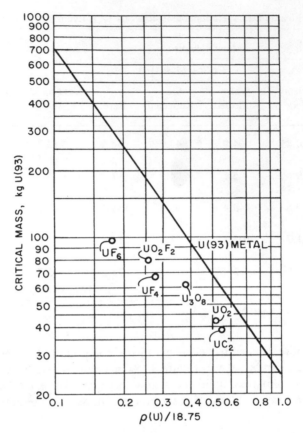

Figure 8.1 The effect of density on the critical mass of 93%-enriched uranium metal and compounds. (From Ref. 8.)

of atoms of moderator (hydrogen) present to the atoms of the fissile (X). Figure 8.2 shows the situation with ^{233}U as an example. (Note that a safety factor has been applied in this case.) As one moves from the pure metal on the right to the left, the moderating effect of the increasing water content causes a drop in the total amount of ^{233}U that can be handled as a single lot. Neutron absorption by the hydrogen is however increasing at the same time and this opposing effect first brings about minima in the curves, then allows the limiting mass to increase. Thus at a concentration of about 60 g/l in the ^{233}U case the amount of the isotope that can be handled is at its smallest value—the 'minimum critical mass.' The allowable total rises rapidly to the left of the minimum point and a concentration is eventually reached where neutron absorption by the hydrogen makes criticality impossible. This is the 'limiting critical concentration.' At any lower concentration the solution is safe no matter how much total material is present. (There have however been criticality accidents that came about when solutions thought to be sufficiently dilute were run into vessels where it was not

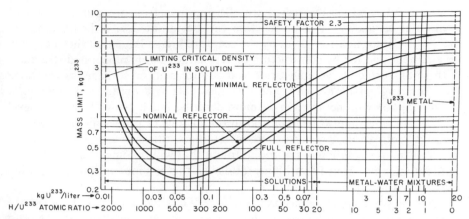

Figure 8.2 Change in the critical mass of ^{233}U with the H/X ratio in aqueous solutions or as metal mixtures with water. (From Ref. 2.)

realized that precipitates were on the bottom from previous operations or a layer of organic solvent present that extracted and concentrated the fissile content.)

Criticality data for solutions, for mixtures of metal with water and for slurries and salts are again usually calculated for the minimal, nominally reflected and wholly reflected situations and for the spherical, cylindrical and slab configurations. Since it is difficult to imagine a 'bare' solution without some sort of container being involved, a reflector of $\frac{1}{16}$–$\frac{1}{8}$ in. thickness of stainless steel or other common metal such as iron, copper, aluminum, nickel or titanium is assumed for the 'minimal' calculations.[2] The water reflector for the nominal and fully reflected situations is assumed to completely surround this metal container.

Use of the infinite-cylinder and infinite-slab data is qualified for solutions. The solutions must be uniform throughout, the reflector cannot be more efficient than an unlimited thickness of water and, in the case of plutonium, the calculations are made on the basis that there will be four nitrate ions present per Pu atom.[3]

The effect of the presence of a moderator in intimate contact with the fissile material is very apparent in Tables 8.1–8.4. The critical masses for the fissiles are in kilograms when the material is as metal, but factors of 10 or more smaller when the fissiles are in solution. This fact has important implications for hot-cell and plutonium glovebox operations with metallic materials. The amount of water or organic solvents in the working area must be very strictly limited or preferably such moderators not allowed at all.

The data in Tables 8.1–8.5 are for 100%-purity isotopes, rarely encountered in practical terms since other isotopes of the element are almost invariably present. Light-water reactor fuels are generally enriched in ^{235}U only to 3%–4% with the balance being essentially all ^{238}U. A less usual reactor fuel is enriched to 93.5% ^{235}U and is obviously more of a problem from the point of view of accidental criticality. A chain reaction can be initiated in ordinary uranium (0.72% ^{235}U) but neutron losses by ^{238}U capture are so large that they can only be compensated for by construction of very large, highly moderated structures with small surface-to-volume ratios to minimize neutron leakage (as was done with the original Fermi pile). The ^{238}U thus acts as a poison and initiating a chain reaction with neutrons in uranium depleted much below the natural level of ^{235}U becomes impossible. Natural uranium rods of proper diameter can be arranged into a very large array and made to go critical. It is however impossible to form any solution of natural uranium that will become critical. The ^{235}U content must be about 1% or higher, and even this assumes nothing present but uranium and water.

A somewhat similar situation exists for ^{239}Pu, although generally not to the same degree. Because the isotope has a relatively large neutron-capture cross section, ^{240}Pu is built up rather rapidly and again acts as a thermal neutron poison. However, it in turn captures a neutron to form ^{241}Pu which is highly fissionable and thus becomes part of the criticality problem. Most reactor fuels of today are 'burned out' only to a limited extent so the ^{241}Pu content is generally low, but should not be ignored. Reactor fuels of the future, particularly if based on ^{239}Pu rather than uranium, will be another problem.

Uranium-233 is not currently used as a fuel, but probably will be in the future since it can be produced from thorium in breeder reactors. Since ^{233}U has a rather low capture cross section to form ^{234}U (49 barns), the isotope can be prepared in quantity in relatively pure isotopic form without having to apply subsequent enrichment techniques.

While ^{240}Pu is a thermal neutron poison, it does fission with good cross section with neutrons of intermediate (resonance) energies. The safest approach is to treat the isotope as if it were ^{239}Pu, that is, add the amount present to the quantity of ^{239}Pu when making criticality judgments. Figure 8.3 shows the effect of increasing ^{240}Pu content on the critical mass of plutonium in solution. The data are for water-moderated spheres and do not take into account that fissionable ^{241}Pu will also increase as the percent of ^{240}Pu in the plutonium increases.

Since the various fissile isotopes have different critical masses the problem of computing the effective total arises for mixtures. The following formula[10] is useful:

$$^{235}\text{U} + 1.6^{239}\text{Pu} + 1.4^{233}\text{U} + 3.1^{241}\text{Pu} = \text{effective total (grams)}.$$

In other words, the indicated multipliers are applied to the number of grams present of each isotope and the sum taken. The amount thus calculated must then be below the predetermined permissible limit for the particular operation being undertaken. The relatively low critical mass of ^{241}Pu is reflected in its high multiplier.

It will be seen from Tables 8.1–8.4 that the fissile isotopes are all odd-even, that is, either the total number of protons or the total of neutrons in the nucleus is an odd number. (A, the atomic weight, minus Z, the atomic number, is odd.)

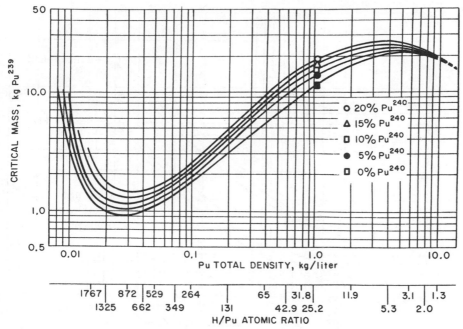

Figure 8.3 The effect of the ^{240}Pu content on the critical mass of plutonium in water solution. (From Ref. 9.)

All of the fissionable but nonfissile species are in general even-even (table 8.5), although it should be noted that the critical mass of ^{238}Pu as metal is comparable to those of the fissiles, but safe in any reasonable concentration in solution. The limiting critical concentration values for the even-evens as shown in Table 8.5 are in terms of kg/l and are all so high that it would be impossible to get that much material into true solution. About the closest one could come in most cases would be to form a low-water-content slurry.

Some other points might be noted concerning the data in Tables 8.1-8.5. A single piece of uranium metal (no interaction in an array) must be enriched in ^{235}U to a minimum of 5% before it can become critical. A simple ^{235}U-H$_2$O mixture—again as a single unit—must be 0.9% enriched (literature values range up to 1.034%). In the much more probable situation, a solution of uranyl nitrate, the enrichment has to be 1.94% to overcome loss of neutrons by absorption in the nitrate. UO$_3$ in homogenous aqueous mixtures must have an enrichment of more than 0.97% and UO$_2$ of more than 0.96% before criticality is possible. (These last numbers were given in a draft version of TID 7016, Rev. 2, but did not appear in the final report.)

Plutonium in true solution is generally as Pu(NO$_3$)$_4$. The effect of neutron losses by absorption in the nitrate can be seen in the data of Table 8.3 and is further demonstrated in Figure 8.4. There will always be some excess nitric acid present when plutonium metal or oxides are dissolved. The curves show

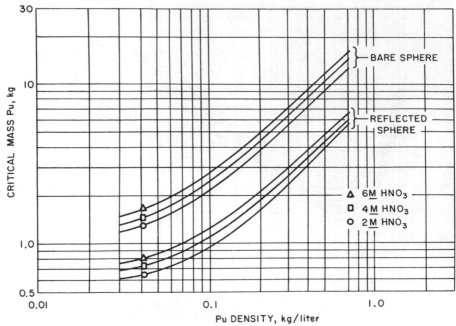

Figure 8.4   The effect of excess nitric acid on the critical mass of plutonium in water solution. (From Ref. 9.)

the effect of this excess acid on the critical mass of ^{239}Pu solutions as a function of the plutonium concentration. The minimum critical mass has still not quite been reached at a concentration of 30 g/l.

It is possible to visualize organic salts of plutonium or uranium that would be soluble in organic solvents, or alternatively, a slurry of metal powder in such a solvent. The H/X ratio could then possibly be higher than in a water solution at the same concentration and the critical mass would be affected accordingly. Thus it has been reported[11] that a Pu-polyethylene mixture has a surprisingly 30% smaller mass limit (0.35 kg) due to the high H/^{239}Pu ratio. This could be of importance in handling plutonium glovebox waste.

The numbers shown in Tables 8.4 and 8.5 are predominantly calculated since sufficient quantities of most of the isotopes for experimental measurement do not exist, although there is a limited amount of such data for ^{238}Pu and ^{244}Cm. In Table 8.4 it will be seen that the calculated critical masses for the heavier fissiles are in the kilogram range as with the lighter nuclides. Putting the heavier species into solution however lowers the critical mass much more drastically than in the case of the lower-weight isotopes.

Intuitively one would think that the higher the fission cross section of a fissile, the lower would be the corresponding critical mass. There is some tendency in that direction, but σ_f is a long way from being a good indicator as is evident below. The capture-to-fission ratio probably plays a more important role.

Isotope	σ_f (barns)	Critical Mass (Reflected solutions, grams)
^{233}U	525	550
^{235}U	577	760
^{239}Pu	810	510
^{241}Pu	1100	260
^{243}Cm	590	213
^{245}Cm	1850	42
^{247}Cm	200	150
^{251}Cm	3000	10

The comparatively low critical masses for even-even ^{238}Pu (as metal) have been pointed out previously. From Table 8.5 it will be seen that the values for metallic ^{244}Cm are also not too different from those of metallic ^{235}U. Clayton and Bierman[5] make an interesting point concerning the even-evens. In a sense, the conditions for criticality are reversed from the odd-even case. Adding moderator to an even-even nuclide makes criticality impossible by reducing the neutron energies down to levels where σ_f is very small. Adding moderator to an odd-even fissile of course has the opposite effect.

8.4 SLURRIES AND SALTS

Slurries are very fine metal dust or powdered compounds suspended in a liquid. If the slurry is uniform throughout the limits given in Tables 8.1–8.5 for solutions can be used for criticality assessments, but with the usual number of qualifications[3]:

1 Each Pu in a plutonium slurry must be intimately associated with four nitrate ions.
2 The H/X ratio cannot be different from that in a true corresponding solution.
3 The ^{240}Pu content must exceed the ^{241}Pu in plutonium-isotope mixtures and the ^{241}Pu weighted and considered as ^{239}Pu.

A true slurry has been defined as one where the individual particles have a surface-to-volume ratio of at least 80 cm.$_{-1}$. Such particles will have diameters of about 0.75 mm.

The cylinder and slab dimensional units given for solutions can also be used for nonuniform slurries provided the conditions given above are met, and that for cylinders the concentration gradient is only along the length and for slabs only parallel to the faces. Reference 3 gives other information on nonuniform slurries.

Table 8.6   Criticality Data for Fissile Compounds

Material Form	Critical Mass (kg)	Full Density (g/cm³)
^{235}U metal[a]	20.1	18.8
UC_2	27.0	11.1
UO_2	29.6	10.8
U_3O_8	43.5	8.3
UF_4	47.9	6.6
UF_6	69.6	4.9
^{239}Pu metal[a]	4.9	19.7
PuO_2	9.0	11.4
Pu_2O_3	9.0	11.4
PuF_3	10.8	9.3
PuF_4	16.0	7.0
$PuCl_3$	36.0	5.7

Source: Reference 2.

[a] For comparison. Compounds are all 100% ^{235}U or ^{239}Pu, fully reflected.

Table 8.6 shows critical-mass data for certain compounds of uranium and plutonium.[2] The values in the table are on the basis of no moderator present but with full reflection. The compound densities are also given. If no moderator is introduced but the full density decreased (as in a heap of pellets) the subcritical-mass limits can be increased by factors taken from Figure 8.5.

Comparison of Figure 8.1 and Table 8.6 shows the effect of lowering the ^{235}U enrichment from 100% to 93%.

8.5   MODERATORS AND REFLECTORS

Because the good moderators are also good reflectors it is easy to confuse the two. The good moderators, those most effective in reducing neutron energies to the thermal range, are actually quite limited in number—light and heavy water (H_2O and D_2O); graphite; beryllium as the metal, oxide or carbide; paraffin; and most plastics. The moderating properties of heavy water are such that it can be used with natural uranium to build smaller reactors, the problem of course being the cost (D_2O is currently $90/lb). Light-water thermal reactors require the use of enriched uranium, which also adds considerably to the cost, but this reactor type predominates because of the ready availability (at least to now) of H_2O which also acts as a coolant as well as a moderator.

The comparative efficiencies of some of the moderators is indicated in Figure 8.6. The data are for unreflected spheres of U(93) diluted with the indicated moderator materials.

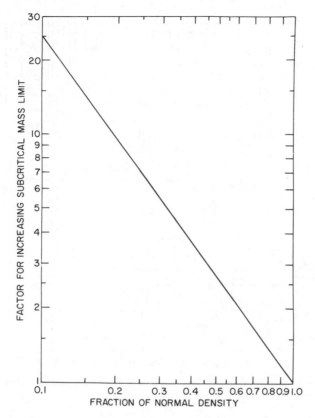

Figure 8.5 Factor by which subcritical mass limits may be increased when fissile material densities are less than normal. (From Ref. 2.)

All materials will act as reflectors to varying degrees, some of them actually being more effective than water, depending on the median energy of the neutrons in the system and the thickness of the reflecting material. This is shown in Figure 8.7 for U(93.5) metal spheres. As can be partially deduced from the figure, a thin stainless-steel containing wall of a vessel will reduce the effectiveness of a water reflector, but a thick layer of steel may be equivalent to, or even better than water.[9] In the case of fast-neutron systems containing even-even fissionables such as ^{237}Np, steel is a substantially more effective reflector than water because there is much less degradation of energy during the reflection process. Water is used as the standard reflector in criticality calculations because it is more likely in practical situations to closely surround a fissile source—as in a reactor.

Concrete as a reflector is not shown on Figure 8.7 but it is equivalent to or somewhat better than water in its effectiveness. It would however be highly unusual for concrete to be fitted closely around a fissile system, and in most practical situations there will be an air gap present. The existence of such a gap between the fissile and nearby walls, etc., has a definite effect in increasing the level of critical mass.

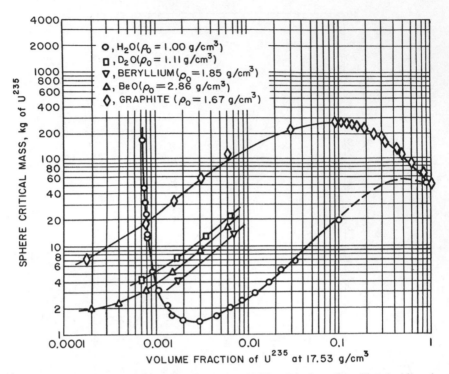

Figure 8.6   Critical masses of unreflected spheres of 93% enriched uranium diluted with various moderating materials. (From Refs. 8 and 12.)

Paraffin and the high C-H plastics are of course effective thermal neutron reflectors.

8.6 POISONS

Any nuclide in or around a fissile system that absorbs neutrons will have the effect of requiring a higher mass of fissile before criticality becomes possible so those isotopes or elements having high neutron absorption cross sections are considered as poisons. Such materials of course play an important role in reactor technology as control rods to maintain subcriticality during shutdown, to control the reactor exactly at criticality during operation, and, when slammed into place by various means, to act as safety devices during emergencies in order to prevent supercriticality.

Cadmium and boron are the two elements usually thought of in connection with poisoning, although some of the rare-earth and other elements also have high σ_c values. Processing equipment and vessels can be made more safe by introducing poisons in strategic locations, the problem then being that of ensuring that the poisoning material stays in place and does not deteriorate with

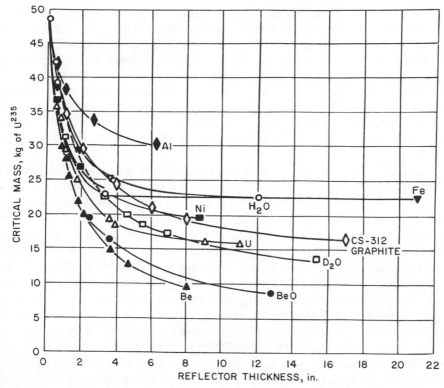

Figure 8.7   Critical masses of 93.5% enriched uranium metal spheres in the presence of various reflectors. (From Refs. 8 and 12.)

time. Beads, Raschig rings[13] or tubing of Pyrex (which is high in boron content) can be placed in process vessels, as an example, but there has to be some means available to be certain that these poisons maintain their integrity indefinitely. Poisons dissolved in solutions are not very satisfactory because of the chance of loss by selective precipitation or plating out on walls. Cadmium foil can be wrapped around a vessel and is very effective in increasing the critical-mass limits of the contents if the cadmium in turn is surrounded by water. The foil by itself, however, will serve as a partial neutron reflector and, depending on the other variables, could actually degrade rather than improve the situation.

8.7 CRITICALITY ACCIDENTS

Accidents involving criticality in this country have been reviewed in detail in several publications[2,14,15] in varying detail. During the lifetime of the AEC (1946–1975) there were twenty-six incidents with none occurring during the last seven years of the agency's existence. During the period the accidents resulted in six deaths and about $4.5 million in property damage. Three of the fatalities and 98% of the money loss occurred in one accident, the explosion at the SL-1, 'Army' reactor at the Idaho Reactor Testing Station in 1961. Three of the power excursions were preplanned, but more energy was released than expected. Nine took place behind heavy shielding, fourteen happened during experiments (mostly on criticality), and six in production or processing facilities. The conditions in each of the last group were somewhat different, but two of the accidents will be described to indicate the ease with which assumed routine operations can lead to unexpected results. The first incident occurred at the Idaho Chemical Plant in 1959—fortunately behind heavy shielding so no one was hurt.

A bank of geometrically safe storage cylinders containing U(93) solution (170 g^{235}U/l) was being air sparged. The sparging apparently initiated a siphoning action that transferred about 200 l of solution into a 5000-gal tank containing about 600 l of water. The resulting supercriticality in this tank resulted in an estimated 4×10^{19} fissions over a period of 20 min, probably following a pattern observed in other accidents. There presumably was a large original 'spike,' following which the system temporarily went subcritical due to the temperature rise and bubble formation causing a decrease in density. As the bubbles disappeared the system again went critical, causing a second spike which again was suppressed by bubble formation. Such cycling went on until there was a more or less stable boiling of the solution, followed by termination of the reaction after some 400 l of water had been distilled into a second tank.

The second accident was much more serious since it resulted in a fatality. The power excursion occurred in some Los Alamos processing facilities in 1958. Six large tanks used for solvent extraction recovery of plutonium and americium from very dilute solutions were being cleaned; and for convenience the residual materials and nitric acid from four of the large units had been run into a single smaller tank. Investigation after the accident showed that this resulted in an 8-in.-thick layer of organic solution containing 3.27 kg of plutonium floating on a dilute aqueous layer containing 60 g of the element. The system at this point was almost at criticality and when the operator approached the tank and turned on a stirrer the mixing of the two layers caused an excursion. The operator was knocked off the ladder he was using in order to look into the tank through a sight hole and ran out of the building. He received an exposure of about 12,000 R and died 35 hr later. It is thought that the excursion produced about 1.5×10^{17} fissions.

In the twenty-six incidents since 1943 the estimated total fissions per event have ranged from 10^{14} to perhaps 4×10^{19}. While direct experimentation to relate these numbers to the dose received by any exposed persons is obviously impossible, a rough estimate can be made based on analysis of the known accidents, particularly one that occurred at the Y-12 facility in Oak Ridge in 1958 and which was duplicated as far as possible in a mock-up at the ORNL Health Physics Research Reactor. The results of this study are given in Figure 8.8.[16,17] It will be seen that for a burst of 10^{18} fissions the total absorbed dose would be well within the lethal range 20 feet away from the unshielded source.

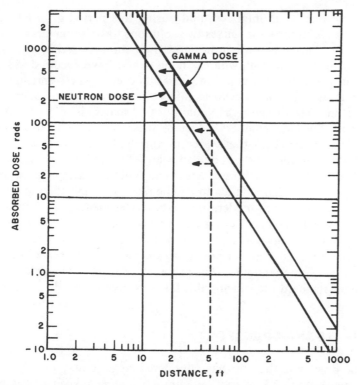

Figure 8.8 Prompt radiation dose from a criticality accident involving 10^{18} fissions. (From Ref. 17.)

8.8 SAFETY FACTORS

There are almost always many variables and uncertainties to deal with in assessing a situation for potential criticality problems, so the need to have safety factors applied to the basic data as given so far is obvious. A factor of 2.3 is frequently applied in the literature for the allowable mass in any solution (the base critical mass divided by 2.3). This figure was derived primarily by consideration of 'double batching.' In other words the allowable critical mass is halved in order to protect against the situation where two lots of material are inadvertently introduced into the same vessel—the extra 0.3 being thrown in to allow for errors in chemical analysis, recordkeeping, etc.. A safety factor of 1.3 is often applied to critical volumes and the 'safe' dimensions of cylinders and slabs. One safety feature that has not been touched upon is the 'double contingency' principle. This simply means that equipment and processes should be designed so that two unlikely events would have to occur simultaneously for criticality to occur. As an example consider a tank designed to be safe with 5%-enriched uranium at full capacity, but also contaning a layer of Pyrex beads sufficiently thick and with enough poisoning effect so that a 10%-enriched solution could actually be tolerated. The beads have been removed for replacement without everyone being informed, and during this period a mistake is made elsewhere in the process that mixes two solutions to bring the uranium enrichment to 10%. This solution is introduced into the supposedly safe tank and causes a criticality accident, whereas none would have occurred if the beads were still in place or if the uranium was at its presumed 5%-enrichment level.

There is also a type of reverse safety factor—conditions allowing some relaxation and permitting higher quantities of materials to be handled than indicated by the base numbers. Most of these favorable factors have been touched upon previously—the degree of enrichment in the case of ^{235}U, the level of ^{240}Pu in plutonium (with due allowance for the ^{241}Pu level), the effect of overall density, the use of poisons, etc. The Nuclear Safety Guide[2] mentions several other factors that would allow some relaxation, including intimate mixture of the fissile with non-moderating materials [elements of atomic number 11 (sodium) or greater]. This of course is the case for many of the salts listed in Table 8.6. The fact that a *thin* layer of steel will decrease the effectiveness of a water moderator (important in case a processing area is unexpectedly flooded) has been previously mentioned. The presence or absence of shielding is also a factor and a standard recognizing this fact has been issued.[18]

8.9 ARRAYS AND TRANSPORT

All of the discussion thus far has considered single units—one fuel rod, one lump of metal, one tank of solution, etc. A whole new layer of complexity is however added when more than one unit of fissile material must be handled, stored or transported in close proximity to others. The neutrons emitted by one can initiate reactions in another, leading to more neutrons overall, and this cyclic pattern of exchange can lead to supercriticality if proper precautions are not taken. One of the touchiest problems is in processing plant design where the effect of each pipe and vessel on others around them must be considered, even if they are individually safe. The normal or possible accidental presence of moderators and reflectors must be taken into account. Even the intrusion of human operators—since the high water content of the body makes it a good moderator and reflector—can be a factor.

A series of articles[19-22] in *Nuclear Technology* gives a thorough survey of the experimental and theoretical studies that have been undertaken dealing with fissile arrays. The first paper reviews the experiments that have been performed; the second discusses the various models that have been used in devising theoretical treatments; the third describes the Monto Carlo approach, one of the better calculational techniques (if you have access to a large computer); and the last applies the different theoretical approaches to a series of hypothetical arrays as a means of determining the preferable method for varying situations. Arrays are also discussed in all of the review-type references previously cited.[2,4,8,9]

The *Nuclear Technology* series concludes[22] that none of the methods satisfactorily handles nonuniformly spaced arrays, arrays with arbitrary amounts of internal moderator or 'clumped' arrays, that is, arrays of arrays. Application of the various calculational methods to a variety of arrays leads to the conclusion[22] that the Monte Carlo approach is best for criticality evaluation of specific problems while the remainder of the models are most useful in establishing possible designs for a fissile storage or processing facility.

Once the various parameters for a storage system are determined—allowable number of units, spacing, amount of fissile per unit, etc.—much can be done by engineering design to assure that each unit remains in its proper place ('birdcaging'). A processing plant is obviously a more complicated problem since the system is dynamic rather than static. Simultaneous transport of several fissile units is again a matter requiring careful calculational analysis and a high degree of sophisticated engineering since adequate shielding and cooling must be furnished as well as a number of safety features to apply in case of an accident.

8.10 REGULATIONS AND ADMINISTRATION

The NRC regulations relating to criticality control appear in 10CFR70.24 and apply to any organization licensed to receive more than 500 g of ^{235}U, 300 g of ^{239}Pu or 300 g of ^{233}U. Requirements for a criticality accident monitoring system and for emergency plans and drills are given. This section of the CFR then serves as a basis for several Regulatory Guides, the most pertinent to the present chapter being No. 3.4.[23] (Nos. 3.33, 3.34 and 3.35 deal with evaluation of the consequences of a criticality accident in a fuel reprocessing plant, a uranium fuel fabrication plant and a plutonium fuel fabrication plant, respectively; No. 3.41 is concerned with the validation of calculational methods; No. 3.43 discusses storage of fissiles; and No. 8.12 gives requirements for criticality accident alarm systems.) Regulatory Guide No. 3.4 is actually very short and in essence simply accepts American Standard ANSI N16.1-1975[3] with the caveat that the standard is not a substitute for detailed criticality safety analyses of specific operations.

This standard is one of several[13,18] developed by Subcommittee 8 of the Standards Committee of the American Nuclear Society working in close collaboration with the American Standards Institute. (Copies of the standards can be obtained from either organization.) The standard presents generalized basic criteria and specific limits for certain single fissionable units but not for multiunit arrays. More detailed guidance is given in ERDAM Chapter 0530[24] and in the Nuclear Safety Guide.[2] These documents are in general consistent with each other since they were largely prepared by or in consultation with the same group of experts.

ERDAM Chapter 0530 first defines responsibilities within the ERDA (now DOE) organization then presents a series of definitions and references. A section then deals with the selection, training and retraining of personnel qualified to handle critical materials and establishes the requirements to be satisfied before an organization can initiate work in the field. The use of a criticality alarm system is specified. Rules are established for conducting independent internal audits of the administrative and physical control techniques used within a facility to ensure criticality safety as are carefully detailed aspects of the review to be considered by the participants. The chapter concludes with a discussion of the various control parameters than can be adjusted for nuclear safety and a brief outline is given of the rules applying to off-site and on-site movement of fissile materials.

8.11   A SUMMATION OF CONTROLS

Table 8.7 is based on Ref. 9 and gives a good outline not only of the hazard-control techniques that can be utilized but also indicates how each category of controls is dependent on the others. Research provides the basic data regarding the fissiles so that parameters such as mass and concentration limits, moderator and reflector effects, etc., can be specified. The engineers then use the data from the scientists to design reaction vessels, piping systems, etc., that are intrinsically safe and to devise monitors, emergency systems and instrumentation to forestall any accident that the designer's ingenuity can foresee as being possible. The administrators—the persons who will operate the facilities constructed for them by the engineers—then take over and establish operational controls, training programs, accountability systems, and so on in order to eliminate as far as possible the factor that frequently is the weakest link in the chain—human error.

Table 8.7   Components of Criticality Hazard Control

Research	Engineering	Administration
Specifications	Equipment	Written procedures
Shape	Design	Approval requirements
Volume	Arrangement	Double signature
Mass	Reliability	Personnel training
Concentration	Instrumentation	Restricted use
Enrichment	In-line monitors	Lockout
Effects	Safety devices	Accountability systems
Moderator	Interlocks	Clear labeling rules
Reflector	Alarms	Clear instructions
Poisons	Vacuum breakers	Visual inspection
Interaction	Filters	Frequent sampling
	Check valves	Double analysis
	Traps	Running inventories
	Overflows	Status records
	Waterproofing	
	Birdcaging	
	Segregation	

Source: Modified from Ref. 9.

Most of the items in the table are straightforward in terms of the material previously presented. Interlocks, vacuum breakers, check valves, filters, traps and overflows are all design options available to the engineer to make certain that material in the system always remains where it is supposed to be and does not accumulate where it is not supposed to be. Waterproofing indicates a plant design that eliminates any possibility that fissile systems could be accidentally flooded and become critical because of the resulting reflector and moderating effects. Under the 'Administrative devices' heading, rules can be established so that no movement of materials can be undertaken without the approval of at least one and perhaps two supervisors having a complete picture of the operation (double signature). 'Lockout' refers to a rigidly enforced rule regulating the permissible quantity of material in a specified area—no more can be introduced once a preset level is reached. Double analysis indicates that chemical analyses of samples taken from key locations in a process must be performed twice, preferably by different individuals and different methods in order to be certain that the analytical data are correct.

The final responsibility for avoiding criticality accidents falls upon the supervisors and operators, the people who actually handle the material. This necessitates constant vigilance and care no matter how routine the operation. These users are obviously helpless however, if the engineers have provided an unsafe design or the scientist has supplied poor basic data. Constant alertness, knowledge and a conservative approach are absolute requirements in busy facilities handling large quantities of fissionable materials. As a further complication, Clayton[25] has pointed out that there are poorly understood 'anomalies of criticality' that must be kept in mind in some cases in addition to the more straightforward considerations outlined in this chapter.

REFERENCES

1 Chapter 4, Ref, 57.

2 J. T. Thomas. Ed., 'Nuclear Safety Guide, TID 7016, Revision 2.' USNRC Report NUREG/CR-0095 (1978); Anon, 'Nuclear Safety Guide.' USAEC Report TID 7016 (Rev. 1) (1961).

3 ANS (Chap. 1, Ref. 12) Subcommittee ANS-8, 'American National Standard for Nuclear Criticality Safety in Operations with Fissionable Materials Outside Reactors.' Standard ANSI N16.1-1975 (1975).

4 R. D. Carter, K. R. Ridgway, G. R. Kiel and W. A. Blykert, 'Criticality Handbook.' USAEC Report ARH-600: Vol. I (Rev.) (1968); Vol. II (Rev.) (1969); Vol. III (1971). (This is a training-reference manual.)

5 E. D. Clayton and S. R. Bierman, 'Criticality Problems of Actinide Elements.' *Actinide Rev.,* 1, 409-432 (1971).

6 S. R. Bierman and E. D. Clayton, 'Criticality of Transuranium Elements—Unmoderated Systems.' *Trans. Am. Nucl. Soc.,* 12 (2), 887-888 (1969).

7 H. K. Clark, 'Critical Masses of Fissile Transplutonium Elements.' *Ibid.,* 12 (2), 886-887 (1969).

8 H. D. Paxton, 'Criticality Control in Operations with Fissile Materials.' USAEC Report LA-3366 (Rev.) (1972).

9 E. D. Clayton and W. A. Reardon, 'Nuclear Safety and Criticality of Plutonium.' In *Plutonium Handbook* (Chap. 6, Ref. 27), Vol. II, pp. 874-919.

10 Harry Bryant, Secy., '186th Meeting, Criticality Hazards Control Committee.' Argonne National Laboratory, August 12, 1976, p. 10.

11 J. K. Thompson, 'Surprisingly Low Minimum Critical Mass of Plutonium-Polyethylene System.' Internal memo. to C. L. Brown, Battelle Pacific Northwest Laboratories, March 3, 1976.

12 H. C. Paxton, J. T. Thomas, Dixon Callahan and E. B. Johnson, 'Critical Dimensions of Systems Containing ^{235}U, ^{239}Pu and ^{233}U.' USAEC Report TID-7028 (1964).

13 ANS (Chap. 1, Ref. 12) Subcommittee ANS-8,'Use of Borosilicate Glass Raschig Rings as a Neutron Absorber.' Standard ANSI N16.4-1971. (1971).

14 USAEC Div. Operational Safety, 'Operational Accidents and Radiation Exposures Within the AEC, 1943-1975.' USAEC Report WASH 1192 (1976).

15 W. R. Stratton, 'A Review of Criticality Accidents.' USAEC Report LA-3611 (1967).

16 G. S. Hurst, R. H. Ritchie and L. C. Emerson, 'Accidental Radiation Excursion at the Oak Ridge Y-12 Project. III. Determination of Radiation Doses.' *Health Phys.*, **2**, 121-133 (1959).

17 H. J. Moe, 'Plutonium Safety Training Course.' USERDA Report ANL 76-30 (1976).

18 ANS (Chap. 1, Ref. 12) Subcommittee ANS-8, 'Criteria for Nuclear Safety Controls in Operations Where Shielding Protects Personnel.' Standard ANSI N16.8-1974 (1974).

19 C. L. Schuske and H. C. Paxton, 'History of Fissile Array Measurements in the United States.' *Nucl. Tech..*, **30**, 101-137 (1976).

20 D. C. Hunt, 'A Review of Nuclear Safety Models Used in Evaluating Arrays of Fissile Material' *Ibid.*, **30**, 138-165 (1976).

21 D. Dickinson and G. E. Whitesides, 'The Monte Carlo Method for Array Criticality Calculations.' *Ibid.*, **30**, 166-189 (1976).

22 D. C. Hunt and D. Dickinson, 'Comparative Calculational Evaluation of Array Criticality Models.' *Ibid.*, **30**, 190-214 (1976).

23 NRC (Chap. 2, Ref. 14), 'Nuclear Criticality Safety in Operations with Fissionable Materials Outside of Reactors.' NRC Reg. Guide 3.4 (2-78) (1978).

24 ERDAM (Chap. 2, Ref. 44), 'Nuclear Criticality Safety.' ERDAM Chapter 0530 (12/76).

25 E. D. Clayton, 'Anomalies of Criticality,' USERDA Report BNWL-SA-4868 (Rev. 3) (1976).

chapter 9

Transportation of Radioactivity

A reasonable set of regulations for safely moving radioactive material from place to place through the usual methods of transportation must take into account a number of variables:

1 Individual nuclides differ markedly in their intrinsic hazard—the estimated toxicities cover a range of 10^8.
2 The radioactivity may be present in very low or very high concentrations and/or in trivial to very high amounts.
3 The physical form of the material may be such as to considerably increase or to decrease the potential hazard in case of a transportation mishap.
4 The material may contain one of the fissionable isotopes, meaning that nuclear criticality as well as the radioactivity potential for damage must be considered.

To these are added the facts that the transfer can be by any of a number of different modes (rail, plane, truck, ship, barge); that agreements must be reached so that materials can be sent to or received from foreign countries; and that protection must be given not only to humans in transit, but also to radiation sensitive items such as high-speed film that might be in the same freight load. It is obvious that one simple set of rules will not suffice to cover all these variables. The regulations that have evolved are accordingly complex and are constantly being modified. The Office of Hazardous Materials of the Department of Transportation (DOT) has published an informal brochure[1] that is of considerable help in threading the maze.

9.1 BACKGROUND

DOT is the key agency responsible for the safety aspects of shipment of radioactive materials in this country. Other groups, particularly the NRC, are involved to varying degrees as described below. The role of IAEA on the international scale will be discussed in a later section.

DOT was established in 1967. Previous to that time the Interstate Commerce Commission (ICC) had jurisdiction over both the safety and economic aspects of the transport of radioactive materials by land. The ICC utilized the Bureau of Explosives (B of E) of the Association of American Railroads as their technical advisor, and for many years shipping containers for radioactivity required B of E certification. The ICC still retains its economic control responsibilities (issuance of operating licenses and control of freight rates), but those involving safety were transferred to DOT. The Bureau of Explosives is thus no longer directly involved in the development or administration of the national safety regulatory program. Certain B of E shipping containers continued to be accepted during the transition period, but these must now also have DOT-NRC approval.

As with the ICC, whose interests were in land transport, the safety aspects of air transport of radioactive materials, previously under the jurisdiction of the Civil Aeronautics Board, were also transferred to DOT. The comparable responsibility for safe transport by water (ship, barge) was also transferred from the Coast Guard. These last two situations were handled for some time by essentially duplicate publication of the DOT regulations in the Federal Aviation Authority's and Coast Guard's sections of the Code of Federal Regulations. In 1976 these CFR sections were consolidated with those of DOT in 49CFR171-189.

The Atomic Energy Act of 1954 gave the AEC safety responsibility for the use of by-product, source and special materials by its licensees, including the transport of those materials. The opportunities for duplication of effort and of conflict between agencies in this situation were apparent. Potential problems were resolved by a 'memorandum of understanding' between the AEC and ICC in 1966, in a revised version between AEC and DOT in 1973, and most recently between NRC and DOT in 1979.[2] NRC in turn passes certain of the authorities on to the Agreement States for regulation of intrastate shipments.

Very small amounts of 'limited' (or 'exempt' or 'small') radioactive materials may be sent through the mails. The U.S. Postal Service of course covers this situation in its own regulations (39CFR123-125).

Table 9.1 summarizes much of what has been given in the above discussion, noting that the FAA and Coast Guard CFR sections are presumably now in with those of DOT.

This rather involved control system appears to work quite well in practice, chiefly because the other agencies and the Agreement States have written their own rules to be consistent with the DOT regulations. The same rules for shippers (packaging, marking, labeling) and for carriers (vehicle placarding, loading, storage, monitoring and accident reporting) thus apply uniformly across the board. The NRC comes into the picture primarily in having the approval authority for shipping containers for large (Type B or greater) quantities of radioactive materials and for containers for shipment of fissiles. DOT carries out the investigation and prepares an appropriate report in the case of an

Table 9.1 U.S. Radioactivity Transport Regulations

Agency	Reference	Title or Content
DOT	49 CFR 171	General Information and Regulations
	49 CFR 172	(Defines articles subject to regulation)
	49 CFR 173	(Shippers responsibilities)
	49 CFR 174	Carriers by Rail Freight
	49 CFR 175	Carriers by Rail Express
	49 CFR 176	(Rail baggage shipment)
	49 CFR 177	(Shipment by public highway)
	49 CFR 178	Shipping Container Specifications
FAA	14 CFR 103	(Shipment by air)
Coast Guard	46 CFR 146	(Shipment by water)
NRC	10 CFR 71	(Packaging and transport of radioactive materials)
Postal Service	39 CFR 123	(Shipment by mail)

Source: Reference 1.

unusual event (accident, leakage, other incident) occurring in transit. If such an event occurs or is suspected at other than in time of transit, NRC conducts the investigation. The two agencies also cooperate in the development of national safety standards and in the review and evaluation of new packaging designs.

The regulations used by the DOT are those evolved since 1959 by the International Atomic Energy Agency. The IAEA pattern was adopted for air shipments by the International Air Transport Association in 1967 and by DOT in 1969. There is some irony in this since the IAEA has now moved on to a different system as will be described later.

9.2 THE DOT REGULATORY FRAMEWORK

The basic DOT pattern of rules is organized to take cognizance of the variables enumerated in the chapter introduction. The wide range of toxicity among the radioisotopes is handled by assignment of each to a 'transport group.' The fact that the material is more hazardous in one physical form than in another is covered by 'normal form' and 'special form' classification. The questions of the quantity that can be shipped at different levels of regulatory control is treated by division of shipments into 'limited,' Type A, Type B and 'large' quantities, each of which categories has its own set of rules, as does movement of fissiles.

The very early IAEA approach was to divide the radioisotopes that might require shipment into high-, moderate-, and low-toxicity categories, that is, three groups. This scheme eventually evolved into the seven 'transport groups' now used by DOT. A list of about 270 radioactive isotopes or materials has been subdivided into these seven groups (Some of the rare-gas isotopes appear in two places and mixed fission products and the ores of U and Th are listed.)

The list in full is available in various U.S. publications[3-6] and in the 1967 edition of IAEA Safety Series No. 6.[7]

Transport Group I includes the highest-toxicity nuclides such as the trans-uranics and the more intensely active alpha isotopes of natural radium, polonium, actinium, thorium and protactinium (^{226}Ra, ^{227}Ac, ^{231}Pa, etc.). Group II contains some of the intermediate specific activity alpha emitters (^{233}U, ^{234}U, etc.), ^{135}Xe, ^{87}Kr, and ^{41}Ar in compressed-gas form; and a few very-high toxicity beta-emitters (^{90}Sr, ^{133}Ba, ^{154}Eu). Group III takes in the very-long-lived alpha sources (^{235}U, ^{238}U, natural U, natural Th) and intermediate toxicity beta iso-topes such as ^{60}Co, ^{126}I and ^{131}I. Practically all of the remaining beta emitters are assigned to Group IV, including such widely used nuclides as ^{14}C, ^{45}Ca, ^{18}F, ^{24}Na, ^{32}P and ^{55}Fe. Group V includes the more hazardous rare-gas isotopes in noncompressed form, and Group VI the other uncompressed rare-gas nuclides. Tritium as a gas, in luminous paint or absorbed on a solid constitutes Group VII. Tritium in the form of water is in Group IV. Numerically, the distribution breaks down as follows:

Group	Members
I	29
II	20
III	67
IV	145
V	5
VI	3
VII	1

It will be seen that well over one-half of the assignments are to Group IV. The classification rules for isotopes not on the list are shown in Table 9.2. There are also rules in 49CFR173.390 for mixtures of isotopes. Mixed fission products fall into Group II.

An isotope in a 'special form' is as a piece of massive solid metal, as an encapsulated high-integrity sealed source or otherwise in a physical condition

Table 9.2 Transport Groups for Unlisted Nuclides

Nuclide Half-life	Transport Group Assignment[a]	
	Atomic Number 1–81	Atomic Number Over 81
Under 1000 days	III	I
1000 days to 10^6 yr	II	I
Over 10^6 yr	III	III

Source: Adapted from 49CFR173.390(b).

[a] Unlisted nuclides cannot be assigned to Groups IV, V, VI or VII.

Table 9.3   Quantity Limits per Package

Transport Group	Quantities		
	Limited (mCi)	Type A (Ci)	Type B[a] (Ci)
I	0.01	0.001	20
II	0.1	0.05	20
III	1	3	200
IV	1	20	200
V	1	20	5,000
VI	1	1,000	50,000
VII	25,000	1,000	50,000
Special form	1	20[b]	5,000

Source: Reference 1.

[a] Higher amounts are "large quantities."

[b] Except for ^{252}Cf where the limit is 2 Ci.

that makes it highly improbable that it will be scattered and lose its identity in the case of a transport accident. 'Normal form' includes everything else—waste in plastic bags, bottled liquids or powders, low-melting, combustible or frangible solids, etc. The criteria for a special form classification are given in 49CFR173.398(a). A qualifying material is assumed to present less potential hazard because of its high-integrity form and generally can be shipped in larger amounts and without regard for the Transport-group classification that it would otherwise have.

The shipping limits per package are given in Table 9.3. The numbers shown are maximum limits, each based on a specific set of requirements for packaging

Table 9.4   Limits for Devices and Bulk Materials

Transport Group	Articles and Devices		Bulk Materials[a] (mCi/g)
	Per Device (Ci)	Per Package (Ci)	
I	0.0001	0.001	0.0001
II	0.001	0.05	0.005
III	0.01	3	0.3
IV	0.05	3	0.3
V	1	1	
VI	1	1	
VII	25	200	
Special form	0.05	20	

Source: Reference 1.

[a] Low-specific-activity materials.

and handling. The 'limited' quantities in the first column can be shipped through the U.S. mails, provided that certain other Postal Service requirements are met. This is also true for the manufactured articles and radioactive devices whose radioactivity limits are shown in Table 9.4 (luminous-dial clocks, smoke detectors, etc.). The limits given in the last column of Table 9.4 are in terms of specific activities rather than total curies and were established primarily for very large bulk shipments such as those of uranium and thorium ores.

The DOT philosophy thus in effect establishes several different sets of regulations, the details of each primarily dependent on the amount of radioactivity to be shipped within the Transport-group framework. As an example, 10 μCi of a Group-I transuranic nuclide can be shipped through the mail if all the rules are followed, and up to 1 mCi shipped by other means under Type A procedures. Anything between that level and 20 Ci requires Type B treatment, and still another set of rules apply if the quantity for shipment is above 20 curies in a single package and thus into the 'large' category. As a result of this approach the terms 'Type A' and 'Type B' are frequently applied to packaging systems, labeling procedures, etc., as well as to quantities of radioactivity.

The Type A concept was established in order to keep the regulations on traffic in lower-activity shipments (estimated in 1974 as being at the level of 800,000 items per year in the United States[1]) reasonably simple. The NRC regulations in 10CFR71 do not even concern themselves with ordinary Type-A transactions, but deal only with Type B, large-quantity and fissile-element transport regulations.

9.3 PACKAGING

The containment system—the packaging—of radioactive material is the base item among DOT's safety controls on shipments. In the agency's terminology, 'package' refers to the final product, including the radioactivity, as given to the carrier for shipment. 'Packaging' applies only to the nonactive shielding and containment barriers around the active material.

The DOT does not specify precise requirements ('the way') for Type A packaging but rather indicates acceptable ('a way') approaches. The package must be adequate to prevent the loss or dispersal of its radioactive contents and to maintain its radiation shielding properties if the unit is subjected to 'normal' conditions of transport as defined in 49CFR173.398(b). These might involve accidental dropping from the tailgate of a truck, being at the bottom of a stack of other packages, being left on a dock in the rain, etc. DOT has listed some off-the-shelf 'Spec-containers' in 48CTR178 that are considered acceptable for such treatment. These containers include fiberboard and wood boxes and steel drums meeting certain construction standards. The agency also describes a performance-based 'DOT Spec 7A, Type A' general package. A potential shipper can compare his proposed containment arrangement to this standard, and if he considers that his plan is equivalent, can proceed without specific regulatory approval of his design. Approval is also not required if the off-the-shelf

'Spec' containers are used. Foreign-made Type A containers, if properly labeled, are also acceptable without specific DOT approval.[1]

Oak Ridge National Laboratory was for some years the major source of
radioisotopes for all uses both in this country and abroad. The packaging
problem naturally was given a good deal of attention. Figure 9.1 shows one of
the ORNL-developed Type-A packaging assemblies for shipment requiring
limited gamma shielding. The radioactive source is encased in an aluminum
can which in turn is enclosed in a lead capsule. This assembly is then sealed
into an ordinary commercial can. The absorbent material around the lead
shield is there partially to resist shock, but also allows liquid to be shipped in
the inner container. Absorbent material sufficient to take up twice the volume

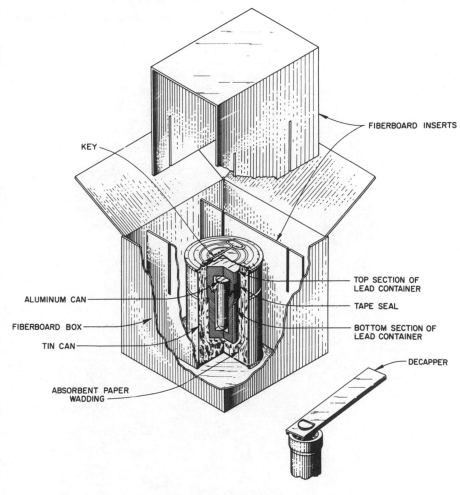

Disposable Container for Solid Materials.
(Lead Shielding – ¼ and ½ inch)

Figure 9.1   Oak Ridge standard packaging for shipment of Type A levels of radioactive solutions.
(ORNL Drawing 16704.)

of the contents is required whenever bottled liquids are transported. The container for the liquid must be both leak and corrosion proof.

Requirements for limited-quantity and Type-A packaging have only been sketched in the above discussion. A variety of tests is required to demonstrate that the package can withstand normal handling conditions.

Type-B packaging, in addition to having to meet the performance standards for normal conditions of transport, must also be able to withstand serious accidents with only limited loss of shielding capability and essentially no loss of containment. The container before being approved must demonstrably be able to survive the following tests without loss of contents:

1. A 30-ft free fall in any orientation onto an unyielding surface.
2. A puncture test consisting of a free drop over 40 in. onto a 6-in.-diam steel pin.

(a)

Figure 9.2 Oak-Ridge developed Type B shipping casks: (a) Finned cask utilizing uranium metal shielding. (b) Lead-shielded cask with fire and impact shield. (ORNL Photos 93395 and 93393.)

3. Exposure in a fire at 1475°F for 30 min.
4. Immersion under 3 ft of water for 8 hr if the unit is to be used for the shipment of fissile materials.

The DOT lists a limited number of specification Type B packagings in their regulations. These designs may be used without need for further approval. Any other design must receive NRC approval before it can be used. The applications for such certifications do not necessarily have to contain proof of actual testing, but engineering calculations, citations of engineering experience, data from model tests, etc., must be supplied in convincing detail. After passing the NRC hurdle, an operating permit request is submitted to DOT. The permit is not granted until other agencies such as FAA and the Coast Guard have also had a chance to review the proposal.

Figure 9.2 shows two ORNL-developed Type-B containers. The first is a cask having depleted uranium metal as shielding which, while more expensive,

(b)

Figure 9.2 (Continued)

allows a reduction in both the size and the weight of the unit. The fins on the sides are to assist in heat dissipation from the radioactive contents and the cask is constructed so that it can be bolted to the floor of the rail car or other transporting vehicle. The second container is a lead-shielded cask which in use is shipped inside the box in the background. The box interior is lined with impact- and fire-resistant materials. (Such casks are also termed 'pots' or, less frequently 'pigs.' The British use 'flasks.')

Large quantities, that is, greater than Type B amounts are usually irradiated reactor fuel elements, materials for establishing gamma-irradiation facilities and presumably will eventually include solidified high-level waste being shipped to disposal sites. The packaging requirements for such materials include all those specified for Type A and Type B quantities plus other provisions for heat dissipation, potential loss of heat transfer media, heavier shielding, and so on. Such shipping systems are of course massive, complex in design and extremely expensive.

Seagren[8] gives a good general discussion of shipping casks as does Langhaar.[9] Oak Ridge has published a 'Cask Designer's Guide'[10] and a series of structural analyses dealing with various aspects of cask design.[11] Chapter 10, Section 10.1 of Volume 3 of the Radiation Shielding Compendium[12] discusses shielding for shipping containers. The AEC periodically issued a directory[13] of packaging containers in use by its contractors, probably a majority of such containers in the country. The American Standards Institute has produced a guide[14] for the design and use of Type A packagings. Division 7 of the NRC Regulatory Guides[15] is devoted to transportation and includes a total of nine different guides, most of which are concerned with packaging.

The steadily growing use of spontaneously fissioning ^{252}Cf as an isotopic neutron source has necessitated construction of some rather unusual shipping containers. These usually come out in a 'fat-man' form having some lead immediately around the source for gamma shielding followed by a large volume of hydrogenous material, usually borated. Figure 9.3 shows a typical container designed for shipment of less than 1 mg of ^{252}Cf.[16]

9.4 HANDLING PROCEDURES

Packages containing exempt or limited amounts of radioactivity cannot have a radiation reading above 0.5 mrem/hr at any accessible point on their surfaces. A warning label is not required on the outside of the package, but the inner container must be marked as containing radioactivity. All packages for larger amounts require exterior warning labels on two opposite sides. These are the familiar diamond-shaped designs with the radioactivity trefoil printed in the upper half. If both halves of the label are white, the surface radiation is less than 0.5 mrem/hr and the package requires no special treatment. If the upper half of the label is bright yellow and the lower portion white with either two or three vertical bars imprinted, a radiation hazard may exist outside the package. The handling rules then become tied into the 'transport index.' This is a con-

Figure 9.3   Cask for shipment of neutron-emitting ^{252}Cf. (From Ref. 16, Courtesy Texaco, Inc.)

cept devised to avoid the situation where several packages, each not particularly hazardous alone, might be placed together in such a way as to constitute a hazard in the aggregate during shipment or storage.

The transport index of a package is defined as the radiation level 3 ft away from the unit's surface, expressed as mrem/hr. The maximum permissible limit for any package containing radioactivity and not shipped in a 'sole-use' mode is 200 mrem/hr radiation level at its surface and 10 mrem/hr at a distance of 3 ft. A package at this maximum limit would then have a transport index of 10 and would require the 'Radioactive-Yellow III' label with three vertical bars in the lower half. (The two bar limits are 10 mrem/hr at the surface, 0.5 mrem/hr at 3 ft.) The total transport index in a single transport vehicle or storage location may not exceed 50, so only five units could be accepted if all were at the maximum level. The DOT regulations specify graded tables of stowage distances versus time for stowage in accordance with the cumulative transport index of a particular shipment.

[A variation of the transport-index (TI) concept is also applied to shipments of fissile materials. Each single package is assigned a fissile transport-index number in accordance with rather complex ground rules, the maximum TI allowed per package again being 10 and the total for all packages in a shipment, 50. Nuclear criticality is of course the concern here, not radiation damage to humans or sensitive freight. If a package has both a radiation and a fissile material TI, the highest value must be taken in summing up the total transport index for the entire shipment.]

The external radiation limits permitted are relaxed for a 'sole-use' shipment, that is, in a closed vehicle (except aircraft) which is used exclusively by a single shipper, loaded by him and unloaded at its destination by the consignee. Under these conditions the radiation reading 3 ft from the external surface of a package can read up to 1000 mrem/hr, but cannot exceed 200 mrem/hr at the external surface of the vehicle, 10 mrem/hr 6 ft away from that surface nor 2 mrem/hr in the driver's cab or any other space occupied by a person. This last requirement may dictate local shielding around the cab or other susceptible areas.

Any rail or highway vehicle carrying materials requiring the 'Radioactive Yellow III' label must also carry external placards indicating a radioactive cargo. Examples of such placards are given in Reference 1.

There are many other restrictions in the regulations. The smallest outside dimension of a radioactive materials package must be 4 in. or greater in order to avoid the possibility of its accidentally or purposely being carried around in a pocket. The maximum surface temperature of the usual package cannot exceed 122°F (180°F for sole-use shipments). As mentioned under the topic of decontamination in Chapter 5, activity removable by a swipe over 200 cm^2 of the package surface must not be more than 220 dis/min of alphas or 2200 dis/min of beta-gamma. The outside of each radioactive materials package must incorporate a seal or its equivalent so that there will be evidence if the unit is illicitly opened. And there are of course extensive paperwork requirements to be met by both the shipper and the carrier.

The shipment of fissile materials involves another set of detailed rules. 49CFR173.396(a) lists some exceptions to the fissile-material shipment regulations, such exempt materials then being treated under the rules applying to all other types of radioactivities. Such exemptions include shipments of not over 15 g of fissile material, uranium or thorium containing less than 0.72% of fissionable nuclide, certain chemical forms of uranium, etc. Regulations applying to the shipment of fissiles appear in various parts of 49CFR171-189, but primarily in Part 173.396.

9.5 IAEA REGULATIONS

The IAEA since 1959 has organized a number of panels and committees and held various symposia to discuss controls on transport of radioactive materials. The history of these activities is discussed at some length by Fairbairn.[17] The seven-Transport-group framework grew out of these efforts, the associated recommendations being periodically published by IAEA through 1967 as revisions of its 'Safety Series No. 6.'[7] The 1973 revision however departs entirely from the Transport Group format, although there are apparently few other major changes in packaging specifications and handling techniques. The U.S.A. DOT has not as yet changed over to the new 'A_1-A_2' concept from the Transport-group approach. The extent to which the new recommendations have been accepted by other nations is not clear, but it is probable that a number of them have made the change since the IAEA has had a particularly strong impact in the transport area. Many countries effectively accept the agency's recommendations and incorporate them into their own laws verbatim.

The reason for the change in the IAEA approach came about primarily because of the 10^8 spread in toxicity of the various nuclides. Division of such a large range into only seven (for all practical purposes, only four) segments still left each with a very considerable difference in potential hazard level between the top and bottom members in a given group. Since the shipping limitations in each Transport group necessarily were based on the most toxic of its nuclides, there were complaints that the restrictions were thus made unnecessarily stringent for many of the others. The A_1-A_2 system responds to those complaints essentially by giving each individual nuclide a transport group of its own.

The revised rules thus specify an A_1 quantity for a given isotope, representing the maximum number of curies that can be shipped in a Type A package in special form. The A_2 value gives the limit when the isotope is in normal form. The two numbers may be the same for some isotopes, but very much different in other cases. The highest A_1 value given for any species is 1000 Ci. There are again special rules for unlisted and mixtures of isotopes, uranium or thorium containing varying amounts of ^{235}U, etc. Any quantities over the A_1-A_2 levels fall into the Type B category and the statement is made: 'The only limits on the activities contained in Type B (U) and Type B (M) packages are those prescribed on their approval certificates."

A Type B(U) package is one that has been unilaterally approved by the competent authority in a single country. Type B (M) packages have received multilateral approvals, that is, by the authorities of more than one country. The amount of activity that can be shipped by mail is expressed as a fraction of the A_1-A_2 limits:

$$\text{solids or gases in special form} - 10^{-4}\ A_1$$

$$\text{solids or gases in normal form} - 10^{-4}\ A_2$$

$$\text{liquids}\quad 10^{-5}\ A_2$$

The definition of the transport index, the permitted radiation level distances, etc., are of course given in IAEA SS No. 6 in metric rather than English measurement units.

The IAEA rules are not simple. The agency presumably received a number of questions following publication of the 1973 revision of the recommendations since a separate Safety Series booklet (No. 37[18]) has been issued to further explain some of the regulations.

REFERENCES

1   U. S. Department of Transportation (DOT), A Review of Transportation Regulations for Transportation of Radioactive Materials.' DOT Office of Hazardous Materials, 400 Seventh St., S.W., Washington D.C. 20590 (issued 1972, revised 1974).

2   Anon, 'Transportation of Radioactive Materials, Memorandum of Understanding.' 44FR38690-38692 July 2, 1979.

3   Anon, 'Hazardous Materials Regulations. Interim Publication.' 49CFR171-177; 41FR 42364-42638 (Sept. 27, 1976).

4   Anon, 'Packaging of Radioactive Material for Transport and Transportation of Radioactive Material Under Certain Conditions.' 10CFR71 (April 30, 1975).

5   ERDAM (Chap. 2, Ref. 44), 'Safety Standards for the Packaging of Fissile and Other Radioactive Materials.' ERDAM Chapter 0529 (12/76).

6   NRC (Chap. 2, Ref. 14) Office of Standards Develop., 'Final Environmental Statement on the Transportation of Radioactive Materials by Air and Other Modes.: USNRC Report NUREG 0170, Vol. 1 (1977).

7   IAEA (Chap. 1, Ref. 2), 'Regulations for the Safe Transport of Radioactive Materials, 1973 Edition.' IAEA Safety Series No. 6 (1973). (Earlier editions were in 1961, 1964 and 1967.)

8   R. D. Seagren, 'Shipping Cask Evaluation.' Isot. Rad. Tech. 8 (3), 350-359 (1971).

9   J. W. Langhaar, 'Casks for Irradiated Fuel: A Look at the Cask Designer's Guide.' Nucl. Safety, 12 (6), 553-561 (1971).

10 L. B. Shappert, 'Cask Designer's Guide.' USAEC Report ORNL-NSIC-68 (1970).

11 J. H. Evans, 'Structural Analysis of Shipping Casks.' USAEC Report ORNL-TM-1312 (Vol. 8 appeared in 1970).

12 J. H. Gillette and J. P. Nichols, 'Shielding of Shipping Containers for Radioactive Sources.' (Chap. 4, Ref. 4, Vol. III, Sec. 10.1, pp. 1-30).

13 Anon, 'Directory of Packaging for Radioactive Materials.' USAEC Report WASH 1279 (1973).

14 ANSI (Chap. 3. Ref. 13), 'American National Standard Guide to Design and Use of Shipping Packages for Type A Quantities of Radioactive Materials.' Standard ANSI N14.7-1975 (1975).

15 NRC (Chap. 2, Ref. 14), 'Division 7 Regulatory Guides: Transportation.'

16 Anon, 'Texaco, Inc., Shipping Containers.' Californium-252 Progress, No. 2 (Jan. 1970), USAEC Savannah River Oper. Office, Aiken, S.C., p. 30.

17 A. Fairbairn, 'Development of the IAEA Regulations for the Safe Transport of Radioactive Materials.' At. Eng. Rev., 11, 843-889 (1973).

18 IAEA (Chap. 1, Ref. 2), 'Advisory Material for the Application of the IAEA Transport' Regulations.: IAEA Safety series No. 37, STI/PUB (1973).

chapter 10

Radioactive Wastes

The emphasis in this chapter will be primarily on the handling of radioactively contaminated solid and liquid waste materials. The discussion will also be chiefly of wastes in the "low-level" category since these are by far of the most concern in laboratory-scale activities.

Gaseous wastes were considered to some extent in the discussion of facility ventilation in Chapter 3. Release of substantial quantities of radioactive gas from laboratory experiments is rather rare, although it can happen. Some method must then be devised either to trap the gas or immediately to dilute it to the point where the emissions from the facility do not exceed the permissible concentration limits given in Appendix B of 10CFR20. Possible approaches for such techniques were also sketched in the earlier discussion and references cited for those seeking more detail.

10.1 CLASSIFICATION OF WASTES

Radioactively contaminated wastes have been classified in a number of ways, with little agreement in the guidelines used by different countries or even between installations in a single country. Categorization into solid, liquid and gaseous forms is generally accepted, but further characterizations vary widely. The system most frequently used is to divide wastes into low-level, intermediate-level and high-level; with further subdivisions such as very low and very high also added in some cases. These are however fuzzy terms, and the lower radioactivity content boundary for high-level liquid wastes as defined in Poland overlaps the upper boundary for low-level materials as defined in Belgium, the USSR, and in several French installations.[1] The U.S. Nuclear Regulatory Commission is attempting to clarify this situation and is sponsoring studies[2,3] aimed at better definitions for use in development of regulations appropriate to the different levels of potential hazard represented by the gamut of possible wastes. These can vary all the way from intensely radioactive spent reactor fuel assemblies to the ash left after burning coals containing traces of natural uranium.

The definition used in the "Glossary of Nuclear Terms" issued by the AEC[4] will be quoted as a first approximation. "Wastes are generally classified as high-level (having radioactivity concentrations of hundreds to thousands of curies per gallon or cubic foot), low-level (in the range of one microcurie per gallon or cubic foot), or intermediate (between these two extremes)." The wastes at any of these levels may also be labeled as TRU or non-TRU, depending on whether or not ^{233}U and/or alpha-emitting transuranic elements are present.

A system for classifying wastes was suggested by the IAEA in 1970.[1] Liquid wastes would be divided into categories one through five, the assignment depending entirely on the activity level per unit volume. Gaseous wastes would be similarly classified but only into three groups, the basis again being the activity level per unit volume (m^3). Solid wastes however would be classified according to the radiation reading in R/hr at the surface of the material. Three categories would be used for solid wastes where the contaminants were predominantly or entirely β emitters, with only insignificant quantities of alphas present. Category 4 would involve the reverse situation—essentially all alpha emitters with very little β-γ present. No category was proposed for packages containing significant levels of both types of activity. This IAEA proposal does not seem to have gained very wide acceptance, at least in the United States.

10.2 LOW-LEVEL WASTES

Low-level wastes are generally considered as being those that can be discharged directly to the environment under controlled conditions after suitable dilution or simple processing. Some of these materials can be disposed of quite simply if the activity levels are sufficiently small. Rules for disposal of liquid wastes into sanitary sewer systems are given in 10CFR20.303. The material must be soluble or readily dispersible in water. The quantity that can be released, ". . . if diluted by the average daily quantity of sewage released into the sewer by the licensee will result in an average concentration equal to . . . [the water concentration limits for occupational exposure as discussed in Chapter 2]." Part 10CFR20 also contains an Appendix C in which quantities expressed in microcuries are listed for the individual isotopes. As an alternate method of control, ten times these Appendix C limits can be disposed of into a sewage system, but the overall concentration averaged over a month's time cannot exceed the MPC for for occupational exposure (taking into account the amount of diluting liquid present during that period). The gross quantity of all radioactive material of any type so released cannot exceed one curie per year per licensee.

Liquid and gaseous discharges to unrestricted areas cannot exceed the applicable MPC limit for nonoccupational exposure via water or air, averaged over a year's time (10CFR 20.106). The licensee can petition for an exception to this ruling by submitting an extremely detailed description of his proposed disposal technique and by giving pertinent chemical, meteorological, topographical, demographic and hydrological data as detailed in the CFR section cited.

Low-level solid wastes can be contaminated paper, wood, ion-exchange resins, small tools, uranium ore mill tailings, biological material, etc. Because of the difficulty of monitoring for low radioactivity in many of these materials all solid waste generated in suspect areas is often treated as it were contaminated. This is a practice to be discouraged since it adds materially to the bulk. A licensee may bury such materials if (1) the amount at burial at any one location and time does not exceed 1000 times the 10CFR20 Appendix C limits, (2) burial is at a minimum depth of 4 ft and (3) successive burials are separated by distances of at least 6 ft and not more than twelve burials are made per year (10CFR20.304). Incineration of solid materials must be specifically authorized by the NRC. Hospitals and research facilities where prompt disposal of contaminated biological materials is highly desirable for sanitary reasons frequently request such authorization. The NRC has prepared a Regulatory Guide[5] to apply when such biological materials must be packaged and shipped off-site, either because of excessive contamination levels or for other reasons.

The amount of contaminated solid waste that can be buried by a licensee is limited and the disposal technique may not be practical in any event. Most low-level solid wastes are accordingly usually shipped off-site for burial elsewhere. Burials were all made on AEC owned property until 1963 when rules were established allowing such facilities to be commercially operated, although still required to be on state or federal lands. A total of six such burial sites have been established, two in the east (New York and South Carolina), two towards the more central part of the country (Kentucky and Illinois) and two in the west (Nevada and Washington). The New York facility stopped operations in 1975 due to leakage problems from the burial trenchs, and the one in Kentucky was closed in 1977. The Illinois area reached its capacity in 1978 and its operators are having political difficulties in acquiring a new license for expansion. The South Carolina facility is nearing capacity and the state has placed limits on the amounts and types of material that can be accepted. The Nevada and Washington areas were until very recently operating normally, but at the time of writing are at least temporarily closed on the basis that some of the materials being received for disposal are sloppily packaged. The low-level waste disposal problem is thus becoming acute, particularly in the northeastern United States where most of the wastes are generated from universities, hospitals, research laboratories and industrial sources as well as from power reactor operations.[6] The problem is of course receiving attention.[7,8]

Five of the six commercial sites are state-regulated through the Agreement-States mechanism, the sixth (Sheffield, Illinois) controlled directly by the NRC. It has been recommended[9] that ownership and control of all radioactive waste disposal sites be transferred to the Department of Energy. That agency presently supervises five major and nine supplementary low-level waste burial sites in connection with its own activities. Addition of the existing commercial locations would allow operation of a national disposal program under a uniform set of practices and criteria.

The burial sites are selected in types of soil where water migration is very slow and where hydrological and geological features are such as to promise maximum retention of the active nuclides. Trenches are dug (300 ft long by 60 ft wide by 25 ft deep being typical.) The wastes, packaged in steel bins or drums, wooden boxes, etc., are loaded into the trench and covered with at least several feet of earth when the excavation is full. There has been evidence of leakage from a few of the trenches at some of the existing burial grounds. Any new facilities will undoubtedly be required to prepare better drainage, containment and monitoring systems and to improve general maintenance procedures before any further operating licenses are granted.

TRU wastes are defined as those contaminated with ^{233}U or transuranics to an average activity of 10 nCi/g or greater. The AEC proposed changes to 10CFR20 in 1974 that would require segregation and retrievable storage of TRU wastes, but these modifications still have not been officially adopted, pending better definition of the overall waste problem. The AEC did incorporate the rules into its own regulations where they appear in ERDAM Chapter 0511 and now apply to DOE facilities and contractors. The user in a DOE laboratory is thus required to segregate the contaminated wastes he generates into combustible TRU, noncombustible TRU and all others (baleable and nonbaleable). The TRU wastes are packaged in metal containers before being placed in a metal bin for shipment to a disposal facility. Here they are stored retrievably so that they can eventually be shipped to a federal depository when such becomes available. Fiberboard drums or TV cartons can be used as the primary containers for non-TRU wastes. These again are packed into large metal bins before shipment.[11]

Reference 6 presents a lengthy table summarizing the requirements established by each of the six commercial burial grounds. All require that liquids be solidified with cement, plaster of Paris or chemicals or by absorption on vermiculite (one site only). Four of the sites will accept liquids under specified conditions and carry out their own solidification procedures for a fee. All solids must be packaged, with items such as filter aids already solidified. Bottled gases are accepted at a few of the sites, with some restrictions. Chemically toxic wastes may not be accepted, at least if the hazard from that source is greater than that from the radioactivity. Charges to the customer are predominantly based on the radioactivity level of the package and its size and there are surcharges for odd shapes, etc. Wastes containing fissiles as usual have their own set of rules.

10.3 INTERMEDIATE-LEVEL WASTES

This is the broadest and most poorly defined waste category. Oak Ridge uses a definition of 10^4 to 10^6 times the MPC values for classification of intermediate liquid wastes.[12] The IAEA[1] made an "unofficial" survey of waste characterization practices ("unofficial" because such limits are not so defined in the regulations of any of the responding countries). A few of the responses for "medium-level" liquid wastes were:

Country	Level (μ/ci/ml)
Poland	10^{-4}–10^{-2}
Sweden	10^{-3}–10^{2}
United States	3×10^{-3}–3×10^{3}
United Kingdom	10^{-2}–10^{3}
U.S.S.R.	10^{-1}–10^{4}

Intermediate-level wastes originate primarily in the nuclear fuel cycle and obviously require more than simple processing before release to the environment. Evaporation, ion-exchange columns, flocculation, carrier precipitation, etc. are among the techniques used to split the liquid wastes into a higher-level concentrate and a bulk low-level fraction. Oak Ridge disposes of some intermediate-level liquid wastes by "shale fracture," a technique borrowed from the oil industry. Wells are drilled into a shale bed or other suitable deeply buried geological formation and lateral cuts made into the sides of the bore by jets of a sand-water mixture of by shaped explosive charges. A mixture of water, waste, and cement is then pumped into the well casing under high pressure and forced into the cracks in the bed where the slurry hardens to effectively become part of the rock itself. Test corings around the pilot plant built to test the method indicate quite satisfactory containment of the wastes.[12]

Solid waste from the fuel cycle is usually placed in steel drums and immobilized with cement at the generating site. Much of the "Radwaste" originating at a reactor site will be of low enough activity content to permit disposal at a commercial burial site. Far more intermediate-level waste will be produced in fuel reprocessing plants, including the cladding "hulls" left behind when the fuels are dissolved, miscellaneous solidified process wastes and highly contaminated hot-cell equipment.

The higher end of the intermediate-level waste category is difficult to distinguish from high-level materials because of the vagueness of the definitions. It is highly probable that the disposal of intermediate materials, particularly if alpha contaminated, will eventually be at a federal repository along with the clearly high-level wastes briefly discussed in the next section.

10.4 HIGH-LEVEL WASTES

Reprocessing of irradiated reactor fuel to recover nonfissioned uranium and the plutonium that has been produced during the irradiation is almost universally by the Purex process. The fuel is first dissolved in nitric acid and the chemical composition of the resulting solution adjusted in accordance with the particular

Purex modification being used. This aqueous phase is then brought into contact (usually in an extraction column) with a nonmiscible organic phase consisting of tributyl phosphate (TBP) in a hydrocarbon solvent. The uranium and plutonium are extracted into the organic phase which is removed for further processing to separate and purify the two elements. The aqueous phase left behind, the "first cycle raffinate," contains by far the bulk of the fission products, any transplutonium elements such as americium and curium produced while in the reactor and small amounts of unextracted uranium and plutonium. This raffinate may contain thousands of curies of activity per gallon, the level depending on many factors such as the length of time the fuel was in the reactor, its original enrichment and the length of cooling time between reactor discharge and the beginning of reprocessing. It is this material that is known as high-level waste and the subject of much current debate as to methods for its safe disposal.

Spent fuel removed from a reactor is transferred for on-site storage into "pools," "ponds," or "canals" where it remains under many feet of water which act as both shielding and coolant. Normally these fuel assemblies would remain at the reactor for periods of months to several years, at which time they would be shipped to a fuel reprocessing plant. There is however currently a ban on any reprocessing of fuels from commercial power facilities which means that some reactors are already being severely strained in terms of available spent-fuel storage space. One solution is to continue to forego reprocessing and not to recover the uranium and plutonium at all. The undissolved fuel then automatically becomes a solid high-level waste and could be shipped in its existing form to a federal repository when one becomes available. It is probably that such material would be stored retrievably rather than disposed of irrevocably in order to allow for the contingency that recovery of the U and Pu in the fuel may become necessary in the future.

Treatment and disposal of high-level wastes involves a number of most interesting technical and political problems and relevant reports and articles on the subject have in recent years have been appearing on the order of 100 per month.[13] It is however a subject somewhat beyond the scope of this book. A few informative references will nevertheless be cited.[14-17]

10.5 WASTE HANDLING

Segregation of wastes into combustible and non-combustible TRU and non-TRU categories was mentioned in an earlier section as a requirement established in many facilities. Such segregation must almost necessarily be at the point of production, both for efficiency and to reduce exposures to operators further down the line involved in disposing of the materials. The person working with the radioactivity is obviously in the best position to assign the wastes to the proper disposal category. The first responsibility is thus his or hers. This individual is also generally best able to estimate the quantity of active material in the waste. This unavoidably involves a certain amount of guessing in many cases, but fortunately the numbers usually only have to be known within reasonable limits.

The TRU versus non-TRU division is not the only one that should be made. The presence of a single small high-level source in a large quantity of low-level material would complicate the handling and disposal problem considerably, as would the inclusion of a needlessly large volume of nonactive trash. There are advantages to be gained in collecting short-lived material separately in a manner so that it can be stored to allow decay reduction of radiation levels. Simple chemical common sense should be used, particularly if liquid wastes are accumulated for a period of time before disposal. Organic solvents should never go in with aqueous solutions, particularly if the latter contain strong acids or oxidizing agents. Similarly, strong acids should not be abruptly mixed with strong bases. Corrosive fluids should be handled with due consideration for the nature of the receiving container. An explosion or even a moderately violent chemical reaction in a waste container holding radioactive liquids is even less desirable than in ordinary laboratory operation.

Segregation of solid low-level radioactive waste in the laboratory is reasonably straightforward. Figure 10.1 shows three of the widely used "Blickman" cans (named after the firm by whom they were introduced). These containers

Figure 10.1 Laboratory solid-waste collection units. ("DAW cans.") (ANL Photo 501-77-33.)

are stainless steel throughout and consist of an inner open-top can, the outer shell and a top containing a sliding section. A fiberboard drum, 1 ft³ in capacity, can be placed in the inner can, or the latter can be lined with a plastic bag. The user steps on the treadle on the front of the can, causing the movable part of the top to slide to one side so that material can be dropped into the inner container. The top automatically recloses when the foot pressure is released. Such cans are routinely surveyed by the health physics group and when full, or if one is found having a high external radiation reading, disposal specialists are called.

The inner drum or bag is removed and replaced. A drum will be capped, the cap sealed in place with tape and the package put into a plastic bag for removal from the laboratory. If a bag was used as the primary container, it is tied off and placed into a secondary bag which is in turn tied off. (These activities are carefully monitored for obvious reasons.)

It will be noted that each of the three cans shown in Figure 10.1 is for a different type of waste. The printing identifying the container for combustible TRU material is in red, that for noncombustible TRU in blue and that for non-TRU in green. The radioactivity trefoil and the cautionary label on the front of each can are in accordance with NRC regulations.

The double-contained trash is transferred from the laboratory to a collection point where the non-TRU materials are placed in large steel bins or sealed into "TV cartons" (heavy cardboard boxes used for shipment of picture tubes) for dispatching to the burial grounds. TRU wastes are similarly packaged, but cardboard containers cannot be used since these materials will be retrievably stored for an indefinite period. Metal containers or heavy wooden boxes are required.

A number of different containers can be used to hold contaminated items of various sizes and shapes. Since a number of such containers may be needed, all of which will eventually be discarded, the effort is made to use commercially available industrial items as much as possible for reasons of economy. Cardboard ice-cream cartons in pint, quart and half-gallon sizes; pressure-lid quart, half-gallon and gallon paint cans; crimped-lid 5- or 10-gal paint cans; the TV cartons; fiberboard drums in various sizes; and 55-gal steel drums all find their uses.

The materials going into these various containers may be at high enough radiation levels to require loading by remote means and shielding during transfer operations. Figure 10.2(a) shows a shielding pot developed for transfer of packaged solids while Figure 10.2(b) pictures several shielded devices for movement of liquid containers of different sizes. A shielded container for solid materials can be improvised by placing a fiberboard cylindrical form in the center of a 55-gal drum, then filling the space around the form and inside the drum wall with concrete. More concrete is poured on the top to form a seal after the inner cavity has been filled with waste. This device would of course be buried in that form. The shielded containers shown in Figure 10.2 are for transfer of materials to accumulation points for further treatment or repacking before disposal.

Figure 10.2 (a) Shielded casks for transfer of solid gamma-active wastes; (b) casks for transfer of liquid wastes. (ANL Photos 501-77-29 and 501-77-30.)

The collection and disposal of liquid wastes is somewhat more complicated than for solids. Small volumes of moderately active solutions can sometimes be absorbed directly into a large quantity of vermiculite, the container then being sealed and subsequently handled as solid waste. Ordinarily, however, the first collection of liquid wastes is in some sort of plastic or glass container. Drain funnels are placed in the floors of the hoods or gloveboxes in some facilities and the wastes collected through tubing into a container below. This vessel may be simply placed in a secondary metal tray for hood work, or be enclosed in a bag-out arrangement for glovebox operations. Light local shielding may be necessary, and extreme care exercised when replacing a full bottle in order to avoid

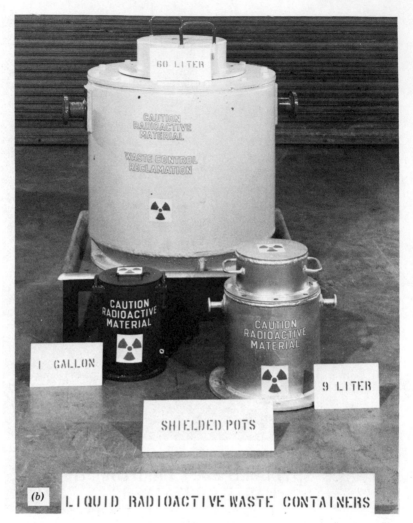

Figure 10.2   (Continued)

spread of contamination. The more usual practice is probably to bottle the wastes directly in the working area. Such filled containers are bagged-out of glove- or containment boxes or dropped into a clean bag if in a hood. The bagged bottle is then placed in a paint can or other suitable container and packed into place with vermiculite or absorbent wadding. These again are operations requiring constant monitoring and exteme care to avoid external contamination on the final package.

The requirement that liquids must be put into solidified form before acceptance at a burial ground presents a problem for the smaller user where the volume of waste produced is high enough to require off-site disposal but not enough to justify construction of special facilities. If the activities being handled are short lived the easiest solution is to provide shielded storage space where the liquids can be kept until their radiation levels are low enough for sewer disposal under NRC rules. In other cases the bulk of the liquid can be decontaminated to a level where it can be similarly discarded, leaving only a small-volume concentrate. Such fractionation can be accomplished by evaporation, ion-exchange resin treatment, use of flocculents or by the use of carrier precipitation methods, a number of which are described in IAEA Safety Series No. 19.[18] The remaining concentrate can be poured into vermiculite (if this is acceptable at the burial ground), or solidified with plaster of Paris, cement, etc., using laboratory-scale equipment. The same procedures will be carried out at a central location in a larger facility, with simple evaporation being the method of choice.

The ICRP has published[19] recommendations for disposal of radioactive materials from hospitals and medical research facilities. The NCRP issued[20] an earlier report on the disposal of ^{32}P and ^{131}I from similar establishments.

10.6 IAEA PUBLICATIONS

The International Atomic Energy Agency periodically convenes panels of specialists to consider various aspects of the radioactive-waste-disposal problem. These meetings usually result in a report appearing as a booklet in either the IAEA Safety or Technical Series. The agency also sponsors or cosponsors international meetings on various aspects of waste handling, the presented papers subsequently appearing in the Proceedings Series. These publications are too numerous for full citation here, but reference will be made to several of those covering more generalized topics.

Safety Series No. 12[21] develops a Code of Practice for disposal of wastes by radioisotope users and No. 19, its Technical Addendum,[18] presents specific information on day-to-day operations and techniques. Safety Series No. 24[22] discusses the various factors that must be taken into account in establishing a waste-disposal system. This report is primarily designed to assist those developing countries where the problem is first being encountered. Other reports in the Safety Series consider disposal into the sea (No. 5), into fresh water (No. 10), into the ground (No.15), and into rivers, lakes and estuaries (No. 36); waste management at nuclear power plants (No. 28); and monitoring of marine radioactivity (No. 11).

Publications in the Technical Series are equally as extensive. No. 116 considers bitumization of wastes. This solidification technique depends on mixing solid or liquid concentrate wastes with asphalt and is widely used in Europe and acceptable at several of the U.S. burial grounds. No. 89 gives details on the chemical precipitation, evaporation and ion-exchange methods used to concentrate activity from bulk liquids. No. 135 describes the use and design of large waste storage tanks. No. 82 considers the treatment of low and intermediate-level wastes. Reference has already been made to No. 101 wherein a waste classification system is proposed. Others could be mentioned.

REFERENCES

1 IAEA (Chap. 1, Ref. 2), Standardization of Radioactive Waste Categories." IAEA Tech Report Series No. 101, STI/DOC/10/101 (1970).

2 J.A. Adam and V.L. Rogers, "A Classification System for Radioactive Waste Disposal—What Goes Where?" USNRC Report NUREG 0456 (1978).

3 G. R. Bray and S. K. Julin, "Final Report on the Classification of Radioactive Wastes." USERDA Report UCRL 13706 (1976).

4 Anon, "Nuclear Terms—A Glossary." USAEC Office of Info. Services (1974).

5 NRC (Chap. 2, Ref. 14), "Packaging and Transport of Radioactively Contaminated Materials." USNRC Reg. Guide 7.2 (6/74) (1974).

6 T.B. Mullarkey, T. L. Jentz, J. M. Connelly and J. P. Kane, "A Survey and Evaluation of Handling and Disposing of Solid Low-Level Nuclear Fuel Cycle Waste." Atomic Industrial Forum, Washington, D.C., Report AIF/NESP-008 (1976).

7 Anon., "NRC Task Force Report on Review of the Federal/State Program for Regulation of Commercial Low-Level Radioactive Waste Burial Grounds." USNRC Report NUREG 0217 (1977).

8 National Academy of Sciences-National Research Council, "The Shallow Land Burial of Low-Level Radioactively Contaminated Solid Waste," (1976).

9 J. M. Deutch, Chrm., "Report of Task Force for Review of Nuclear Waste Management. Draft." USDOE Report DOE/ER-004D (1978).

10 39FR32921 (1974).

11 ERDAM (Chap. 2, Ref. 44), "Radioactive Waste Management." ERDAM Chapter 0511 (9/73).

12 Anon, "Management of Intermediate Level Wastes at Oak Ridge National Laboratory." USERDA Report 1553 (1977).

13 Anon, "Radioactive Waste Processing and Disposal." USERDA Tech. Info Center Report TID-3311-S7 (1976), (Lists 1841 titles appearing Jan. 1, 1975 to July 1, 1976).

14 A. S. Kubo and D. J. Rose, "Disposal of Nuclear Wastes." Science, 182, 1205-1211 (1973).

15 R. A. Kerr, "Nuclear Waste Disposal: Alternatives to Solidification in Glass Proposed." Ibid., 204, 289-291 (1979).

16 H. Krugman and F. Von Hippel, "Radioactive Wastes: A Comparison of U.S. Military and Civilian Inventories." ibid, 197, 883-884 (1977).

17 National Academy Engineering-National Academy Sciences, "Solidification of High-Level Radioactive Wastes, Final Report." USNRC Report NUREG/CR-0895 (1979).

18 IAEA (Chap. 1, Ref. 2), "The Management of Radioactive Wastes Produced by Isotope Users, Technical Addendum." IAEA Safety Series No. 19, STI/PUB/119 (1966).

19 ICRP (Chap. 2, Ref. 31), "Report of Committee V on the Handling of Radioactive Materials in Hospitals and Medical Research Establishments." ICRP Publication 5, Pergamon, Oxford (1965).

20 NCRP (Chap. 2, Ref. 7), "Recommendations for Waste Disposal of Phosphorus-32 and Iodine-131 for Medical Users." USNBS Handbook (1951)."

21 IAEA (Chap. 1, Ref. 2), "The Management of Radioactive Wastes Produced by Radioisotope Users." IAEA Safety Series No. 12, STI/PUB/87 (1965).

22 IAEA, "Basic Factors for the Treatment and Disposal of Radioactive Wastes." IAEA Safety Series No. 24, STI/PUB/170 (1967).

Bibliography and Abbreviations

1 BIBLIOGRAPHY

1   C. M. Lederer and V. S. Shirley, Eds., *Table of Isotopes, Seventh Edition.* Wiley-Interscience, New York (1978).

2   T. W. Burrows and N. E. Holden, "Source List for Nuclear Data, Second Edition." USERDA Report BNL-NCS-50702 (1977).

3   Anon., "USA Standard Glossary of Terms in Nuclear Science and Technology." USAS N 1.1-1967 (revised edition), USA Standards Institute, 10 E. 40th St., New York, NY 10016 (1967).

4   IAEA (Chap. 1, Ref. 2), "International Directory of Certified Radioactive Materials." IAEA Publication STI/PUB 398 (1975).

5   K. Z. Morgan, *Principles of Radiation Protection, a Textbook of Health Physics.* Wiley, New York (1967)

6   J. J. Fitzgerald, G. L. Brownell and P. S. Mahoney, *Mathematical Theory of Dosimetry.* Gordon and Breach, New York (1967).

7   H. W. Patterson, R. H. Thomas and R. Wallace, "Accelerator Health Physics." USAEC Report LBL-900 (1972).

8   ICRU (Chapter 2, Ref. 24), "Radiation Protection Instrumentation and its Application." ICRU Report 20 (1971).

9   S. Glasstone, *Sourcebook on Atomic Energy, Third Edition.* Van Nostrand, Princeton, N.J. (1967).

10 C. H. Koontz, Ed., "Shelter Design and Analysis. Vol. 1, Fallout Radiation Shielding." Office of Civil Defense Report TR-20 (revised edition) (1970).

11 R. D. O'Dell, "Nuclear Criticality Safety." USAEC Report TID 26286 (1974).

12 DOE (Chap. 4, Ref. 1), "Everything You Always Wanted to Know about Shipping High-Level Nuclear Wastes." USDOE Report DOE/EV-0003 (1978).

13 P. F. Rose and T. W. Burrows, "ENDF/B Fission Product Decay Data." USERDA Reports BNL-NCS-50545 (Vol. 1) and BNL-NCS-50545 (Vol. 2) (1976).

14 IAEA (Chap. 1, Ref. 2), "Handbook of Nuclear Activation Cross Sections." IAEA Tech. Report Series 156 (1974).

15　F. K. McGowan and W. T. Milner, "Reaction List for Charged-Particle-Induced Nuclear Reactions." Nucl. Data Tables A **18** (1), 10136 (1976).

16　S. A. Lis, Ph. K. Hopke and J. L. Fasching, "Gamma-Ray Tables for Neutron, Fast Neutron and Photon Activation Analysis." *J. Radioanal. Chem.,* **24** (1) 125–251 (1975); *ibid.* **25** (2) 303–428 (1975).

17　H. W. Kirby, "Decay and Growth Tables for the Naturally Occurring Radioactive Series (revised)." USAEC Report MLM 1042 (1973).

18　Merrill Eisenbud, *Environmental Radioactivity, Second Edition.* Academic Press, New York (1973).

2 ABBREVIATIONS

The various abbreviations and acronyms encountered in this book are summarized in Table A.1.

Table A.1　List of Abbreviations

A-E	Architect-Engineer
AEC	(U.S.) Atomic Energy Commission
ALARA	As Low As Reasonably Achievable
ANL	Argonne National Laboratory
ANS	American Nuclear Society
ANSI	American National Standards Institute
ASHRAE	American Society for Heating, Refrigeration and Air Conditioning
BEIR	Advisory Committee on the Biological Effects of Ionizing Radiation of the National Academy of Sciences – National Research Council
BNL	Brookhaven National Laboratory
CFR	Code of Federal Regulations
DOE	(U.S.) Department of Energy
DOT	(U.S.) Department of Transportation
EDTA	Ethylenediaminetetraacetic acid
EERA	ERDA Energy Research Abstracts
EPA	(U.S.) Environmental Protection Agency
ERA	Energy Research Abstracts
ERDA	(U.S.) Energy Research and Development Administration
ERDAM	ERDA Manual
FDA	(U.S.) Food and Drug Administration
FR	Federal Register
HEPA	High Efficiency Particulate – Air (filters)
HEW	(U.S.) Department of Health, Education and Welfare
HFEF	Hot Fuels Examination Facility

Table A.1 (continued)

HVAC	Heating, Ventilation and Air Conditioning
HVL	Half Value Layer
IAEA	International Atomic Energy Agency
ICRP	International Commission on Radiation Protection
ICRU	International Commission on Radiation Units and Measurement
INIS	International Nuclear Information System
ISO	International Organization for Standardization
LASL	Los Alamos Scientific Laboratory
LBL	Lawrence Berkeley Laboratory
LET	Linear Energy Transfer
LLL	Lawrence Livermore Laboratory
MPC	Maximum Permissible Concentration
MPD	Maximum Permissible Dose
NAE	National Academy of Engineering
NAS-NRC	National Academy of Sciences – National Research Council
NBS	National Bureau of Standards
NCRP	National Council on Radiation Protection and Measurements
NFPA	National Fire Protection Association
NIOSH	National Institute for Occupational Safety and Health
NRC	(U.S.) Nuclear Regulatory Commission
NSA	Nuclear Science Abstracts
ORNL	Oak Ridge National Laboratory
OSHA	(U.S.) Occupational Health and Safety Administration
PHS	(U.S.) Public Health Service
PNL	Battelle Pacific-Northwest Laboratory
RBE	Relative Biological Effectiveness
REIC	Radiation Effects Information Center
RHSO	Radiological Health and Safety Officer
RSTD	ANS Remote Systems Technology Division
SI	International Standard of Units
SRL	Savannah River Laboratory
SRP	Savannah River Plant
TBP	Tri Butyl Phosphate
TIC	DOE Technical Information Center
TRU	Transuranic (contaminated wastes)
TVL	Tenth-Value Layer
UNSCEAR	United Nations Scientific Committee on the Effects of Atomic Radiation

Index